ENVIRONMENTAL SUCCESS STORIES

ENVIRONMENTAL SUCCESS STORIES

Solving Major Ecological Problems and Confronting Climate Change

FRANK M. DUNNIVANT

Columbia University Press
New York

i

Columbia University Press
Publishers Since 1893
New York   Chichester, West Sussex
cup.columbia.edu
Copyright © 2017 Columbia University Press
All rights reserved

Library of Congress Cataloging-in-Publication Data

Names: Dunnivant, Frank M., author.
Title: Environmental success stories : solving major ecological problems
and confronting climate change / Frank M. Dunnivant.
Description: New York : Columbia University Press, [2017] | Includes
bibliographical references and index.
Identifiers: LCCN 2016032434 | ISBN 9780231179188 (cloth : acid-free paper) |
ISBN 9780231179195 (pbk. : acid-free paper) | ISBN 9780231542906 (e-book)
Subjects: LCSH: Environmental protection—Technological innovations. | Green
technology. | Environmental risk assessment. | Climatic changes—Prevention. |
Climatic changes—Risk management.
Classification: LCC TD170.2 .D86 2017 | DDC 363.7—dc23
LC record available at https://lccn.loc.gov/2016032434

Columbia University Press books are printed on permanent and durable acid-free paper.
Printed in the United States of America

COVER DESIGN: Martin N. Hinze
COVER IMAGES: © iStock and Shutterstock

To Lukas and Marley,
may my generation leave your generation a better Earth

Contents

Author's Note   ix

Acknowledgments   xi

INTRODUCTION

ONE

Securing Safe, Inexpensive Drinking Water   7

TWO

Effective Treatment of Our Wastewaters   30

THREE

The Removal of Anthropogenic Lead, and Soon Mercury,
from Our Environment   59

FOUR

Elimination of Chlorinated Hydrocarbons from
Our Environment   80

FIVE
The Safety of Chemicals in Our Food and Water
Risk Assessment 102

SIX
Saving Our Atmosphere for Our Children 114

SEVEN
Legislating Industry
The Need and the Success 129

EIGHT
The Rapid Advancement of Technology
Our Best Hope 140

NINE
Humans' Greatest Challenge
Climate Change 149

TEN
Conclusion and Transition to a Bright Future 170

AFTERWORD
Imagination, Responsibility and Climate Change
KARI MARIE NORGAARD 172

Bibliography 189

Index 233

Author's Note

To the teacher and student:

This book has evolved over the past 20 years as I have struggled to teach environmental issues to undergraduate humanities and social science students in need of a science distribution course and to undergraduate and graduate science students needing a broader understanding of environmental issues. For the humanities and social science students I use this book as my primary text and supplement it with charts, figures, data, and news stories from the Internet. Undergraduate and graduate science and engineering majors learn broader issues of environmental science when this book is used as a supplemental text for discussion. This book is meant to be a living text and my supporting web page at https://sites .google.com/a/whitman.edu/environmentalsuccessstories/home contains news stories that are updated almost daily on Air/Food/Water, China, Climate Change/Climate Chaos, Coal, Politics, Renewable Energy, and Technology. If teachers wish to use my PowerPoint slides they are available from my home page at http://people.whitman.edu/~dunnivfm. We can learn from our successful environmental legacy.

To the popular science reader:

My primary intent for writing this book is education, inside and outside of academia. One of my greatest teenage influences and motivations for becoming a scientist and teacher was popular science books. In this book I attempt to explain the science behind complicated environmental concepts that are only briefly addressed by newscasts. The key to maintaining sustainable conditions on Earth for human society, and other species, is education, and I hope this book will play a part in our fight to convince our industry leaders and politicians to greatly reduce fossil fuel emissions, develop renewable energy sources, and develop sustainable practices worldwide. We can learn from our many past environmental successes.

Acknowledgments

Many colleagues and friends have positively influenced this book. My greatest gratitude goes to Nora Hawkins, who turned my first draft of dyslexic thoughts into a manuscript that others could appreciate and improve upon. My first reviewers were Dr. Marion Gotz, Dr. Leena Knight, and Dr. Dan Burgard, who provided many great ideas. A special thank you goes to Patrick Fitzgerald of Columbia University Press, who recognized the important messages in the book and sought out two excellent readers-reviewers who significantly improved the manuscript. And I again offer my gratitude to Dr. Kari Norgaard for the afterword illustrating the need of social scientists to implement the ideas of natural scientists into society. As always, I owe much of my success to the many students who I have taught at Whitman College since 1999.

ENVIRONMENTAL SUCCESS STORIES

Introduction

It took me many years to realize it, but I have been an environmentalist all of my life. I was raised on a farm in a family of woodsmen. My early years were spent hunting, fishing, and camping, but in my later teenage years I quickly turned to only observing nature, without physically capturing or harming it. I spent my college years in search of a science career, and when I transferred from a community college to a four-year university (Auburn University) I decided to study environmental health, a somewhat obscure college major that I had never heard of before. This program allowed me to study all of my interests: biology, chemistry, and engineering. Friends and relatives asked me the typical question one would ask any college student, "What are you studying?" After hearing my answer, many responded, "How could you possibly make a career out of that?" In the mid-1970s, environmental issues were thought of as a hippy fad, which was expected to fade quickly. I, on the other hand, saw environmental research as a fulfilling lifelong career, and love. I continued to study environmental issues in a Master's of Science program followed by earning a Ph.D. in environmental engineering (Clemson University) and a career working on environmental issues. As we now know, the environmental field and public interest in the environment has grown steadily and intersects with all aspects of life today.

I have lived through most of the environmental movement, and during the early years I was very impressed with societal demands and government action, but after the environmental enthusiasm of the 1970s, the Reagan antienvironmental movement gained momentum in our country. I comforted myself by believing that this was as bad as it would get. Now looking back on the 1980s, I and many other scientists realize that this was the last point at which the U.S. government could have made a difference with respect to global warming/climate change/climate chaos. Reagan did in fact create environmental change, but rather than saving the environment he caused more damage. Unfortunately, a series of do-nothing environmental presidents and Congresses followed. During the George W. Bush years, environmental protection suffered its greatest. As mass hysteria ruled in environmental circles, I was pondering the future with a few fellow scientists and friends, and I realized that Bush slowed down the process of creating beneficial environmental policies, but ultimately, he did not actively inflict any damages to the environment (Schnoor 2004a, 2004b). Like his predecessors, he delayed action on global warming, but industry officials are not ignorant. They knew that the environmental and political pendulum would swing in the other direction, so in general they maintained the status quo with respect to environmental standards.

Around 2004, in a state of reflection, I realized how far the environmental movement had actually come since the 1970s. In the United States (and many other countries) we have made considerable progress with respect to many major issues. As described in this book, we have had great successes in improving our water quality by advancing our domestic and industrial waste treatment and by eradicating waterborne pathogens. Likewise, after decades of abuse by a few bad actors in industry, we have virtually eliminated the toxic metal lead from our environment, and the removal of mercury is well on its way to fruition thanks to action by the Obama administration. In addition, our approach to farm and pest management has greatly been modified with the removal of chlorinated pesticides such as DDT from the market. And we now have very valid and constantly improving methods for risk assessment that balance the beneficial use of chemicals in our modern society with the health effects posed by these same chemicals. Governments have finally acknowledged the destructive impact our overpopulated species can have on large global systems, such as the atmosphere. And developed countries have well-established legal structures in place for addressing environmental issues.

These are the environmental success stories presented in this book. These stories shed light on the major "lessons learned" from our previous successes. We can learn from our history by reflecting on all of the positive environmental changes we have accomplished since the 1970s. We can renew our confidence in our ability to solve global warming/ climate change/climate chaos. All of the environmental success stories covered in this book required monumental effort. The only question that remains is, WHEN will we start to really fight the causes of global warming? The Obama administration started with a mixed record with regard to environmental policies, but his second term in office showed much improvement in this area. We are drilling for and producing more oil and digging up more coal than ever in the United States, we may allow the transport of relatively dirty tar sands through the United States, and we are increasing hydraulic fracking for natural gas each year, yet, as I am in the process of writing this book, President Obama has finally announced that he intends to address climate change by targeting coal-fired plants and CO_2 emissions. We will all be paying close attention to how diligent his efforts are in addressing climate change on the national and international scale.

Residents in many countries have spoken clearly about permanent and effective global environmental polices. This book shows, case by case, what can be accomplished when citizens, governments, and industry work together. While we have made great strides in public health and Planet Earth health, we still have work to do, as noted at the end of each chapter. I will begin with the environmental movement's greatest success story in chapters 1 and 2: clean drinking water and sanitation in developed countries and how it is essential that the technological advances are extended to developing countries. Chapter 3 will delineate the successes in the removal of the toxic metals lead and mercury from our daily lives. Chapter 4 will focus on toxic organic pollutants, specifically the end of the production of chlorinated pesticides such as DDT, and our current and future pollutants of concern, endocrine-disrupting chemicals. In chapter 5, I will describe how risk assessment contributed to environmental success by teaching us when to act in a cost-effective way against a specific chemical based on toxicological data. Chapter 6 will turn our attention to our delicate atmosphere, in particular our great successes concerning smog reduction, stratospheric ozone depletion, and acid rain. Our relatively effective national legal structure and how environmental laws should be implemented worldwide will be the topic of chapter 7. In chapter 8, my goal is to convince the reader that in order to protect our environment in the long term; our first political, social, and scientific steps will need to be conservation, population reduction, and technology development. While we wait to attain a sustainable population while raising the standard of living of others, technology that remedies the impacts of our oversized population can assist this transition. In the final chapter, I will make a case for humanity's greatest challenge, global warming/climate change/climate chaos, and how we can mediate its causes. The major lesson from this book is that for a dedicated population of concerned citizens no problem is unsolvable, including global warming. And in an afterword, Dr. Kari Marie Norgaard discusses imagination, responsibility, and climate change and how social scientists must now move the environmental movement forward.

I hope you enjoy reading about our most impressive environmental successes.

As to the geographic scope of this book: while this book stems from how we have solved and are working to address environmental issues in the United States, it also offers numerous examples of approaches taken in other countries. In particular, a number of environmental problems that have been largely solved in what is called the Global North (previously called developed countries), such as clean drinking water, but remain a critical issue in the Global South (formerly referred to as the developing world). In reference to distinguishing between nations, I will use the currently politically accepted terms Global North and Global South. I completely recognize that there are no acceptable terms to distinguish between the economically "haves" and "have-nots" of our world, but for purposes of discussion, I need to differentiate between these based on current environmental conditions.

More and more the environmental problems that we face today are global in scope, climate change being the clearest example to date. Environmental dilemmas such as ozone depletion have previously been addressed on the international stage and this international collective action will be increasingly required as globalization continues. There is much pessimism about the United States' environmental performance, especially in terms of the willingness of our elected leaders to engage in international collaboration. This is illustrated by the fact that the United States has not signed a treaty on climate change since 1992. There is much that we have achieved domestically in terms of environmental performance, however, and the successes iterated here give us hope for cautious optimism that we can once again come together at an international level to solve the issues facing our planet in the twenty-first century.

I will start this book with my closing statement in chapter 9: if your politicians cannot be educated or do not believe in science, then vote them out of office. This is how the world will change.

ONE

Securing Safe, Inexpensive Drinking Water

Water,

wasser,

voda,

eau,

aqua,

acqua,

hydro,

maya

Through the history of literature, the guy who poisons the
well has been the worst of all villains.

—AUTHOR UNKNOWN

The noblest of the elements is water.

—PINDAR

This chapter documents one of human's greatest success stories, acquiring safe drinking water at relatively low costs and the near elimination of waterborne diseases and pollutants in drinking water systems in Global North and many Global South countries. The astonishing simplicity of water treatment will be presented.

Access to safe, clean drinking water is globally viewed as a basic human right. Several constant themes include: (1) the presence of natural microbial pollution in all waters, (2) that we all live downstream, meaning that water that is one person's waste can become another person's source, and (3) that any discussion of safe, clean drinking water is inseparable from adequate and effective wastewater treatment, the subject of chapter 2. Here we will mention but not focus on securing drinking water, one of our great future challenges. For example, it is estimated that by 2025 to 2030 some Global South countries water needs will be twice their actual supply and that more than half of the world population will be facing water-based vulnerability. Existing water shortages will only be worsened by increasing climate change, the subject of chapter 6. Securing access to clean drinking water is fundamental to human existence; this is an area in which the Global North countries, and soon, much of the Global South, have largely been successful. We will close the chapter with an assessment of our success and what is needed for the future.

Humanity's Fundamental Dependence on the Unique Element of Water

Water and life go hand in hand, no life can exist without water, be it the lowly microbes or the largest mammals or trees. Water covers 71% of the Earth's surface, but only 2.5% of this is freshwater. Of the freshwater that we use in the United States (excluding water in glaciers and water too deep in the ground to access), 70% supports agriculture. Water is considered the medium of life, so much so that it is an essential criterion astronomers look for in the quest for extraterrestrial life. Water also plays an immense role in economics, with a strong correlation between access to safe drinking water and gross domestic product (GDP) per capita. Lastly, one of the most fascinating cycles on Earth, which is fundamental to life, is the sun-driven hydrological cycle of evaporation and precipitation.

The Chemical Abstracts Service (CAS) Registry, the international database of chemical information, identifies more than 70 million organic and inorganic substances. Unique among these is water, H_2O. Water has at least 38 unique properties, properties that would not be expected based solely on its chemical structure. Whether or not we are always aware of it, we interact with water's unique properties on a daily basis, and our livelihoods are dependent on them. We physically experience water's unique ability to absorb heat and regulate our local and global weather patterns. As children, we were awed by water's surface tension when we watched insects walk across it. The fact that water can exist in all three phases (vapor, liquid, and solid) gives us many of our most iconic sites, from the oceans to the Arctic. Other familiar properties of water include its density as compared with other chemicals and how this density changes with temperature increases, its unusually high boiling point, and its unusually low freezing point, resulting in a wide range of temperatures in which liquid water can exist and control our local weather patterns. A unique feature of water is that its solid form, ice, floats on the liquid form; no other chemical behaves like this. Imagine how different aquatic life and the winter season would be if ice formed on the bottom of lakes instead of the top!

Water owes its unique properties to its chemical makeup and size. A water molecule consists of two hydrogen atoms bonded to one oxygen atom. The oxygen atom has two extra pairs of electrons on opposing sides to the hydrogen atoms.

The shape and size of the water molecule allows the hydrogen atoms on one water molecule to interact with the electrons of another molecule. This form of interaction is referred to by chemists as hydrogen bonding and is unique to only three elements: oxygen, nitrogen, and fluorine. The "extra" uniqueness of water is that its small size allows more molecules of water to interact and fit in a three-dimensional lattice than other chemicals. This attraction between water molecules produces its unique properties.

While the chemical formula for water was discovered only in 1805, human's interactions and dependence on water goes back to the earliest of times. Migration routes around the world and early civilizations had a close connection with surface bodies of freshwater. During our nomadic existence, water pollution was not a big problem because tribes could always move on to new, and hopefully clean, sources of drinking water. As permanent civilizations developed, however, almost always near fresh, flowing bodies of water, the problem of pollution became imminent. We all live downstream. In permanent civilizations, one tended to bathe and dispose of waste downstream of one's residential drinking water supply. But what is downstream for one person is upstream for another, creating the problem of surface water pollution. Just think of how many towns are located along the Mississippi River and how many people have withdrawn water, used it, and put it back in the river.

Sources of Water Pollution

But what is pollution? Is it natural or human-made? Is it isolated to Global North countries? Is it organic, inorganic, or biological? Is pollution

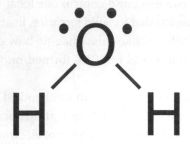

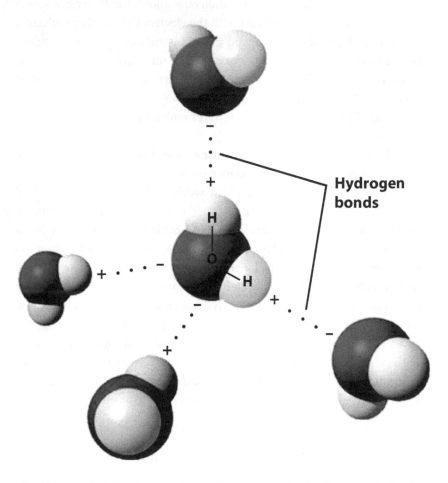

Hydrogen bonds

FIGURE 1.1. (a) Lewis dot structure of water. "Chemistry: Molecular Geometry," in *Structural Biochemistry* (https://en.wikibooks.org/wiki/Structural_Biochemistry/Molecular_Geometry); (b) Hydrogen bonding. "Chemistry: Chemical Bonding—Hydrogen Bonds," in *Structural Biochemistry* (https://en.wikibooks.org/wiki/Structural_Biochemistry/Chemical_Bonding/Hydrogen_bonds).

everywhere? Does it really matter, or is pollution a product of eco-hippy, left-wing rants that undermine capitalism? My education and life are products of the 1960s and 1970s when antipollution efforts were in their infancy, especially from a legal or regulation standpoint. I remember my ecology professor lecturing about how because of pollution, one could no longer drink out of streams in the country without experiencing adverse effects. This was a bit of an overstatement. I continued my education in chemistry, biology, and toxicology, and soon I realized that it would be very unwise to ever drink directly from any stream in the world—today, during the 1970s, or during prior human occupancy of the Americas—because of naturally occurring bacteria. While the problem has gotten worse with human waste in streams, humans, as well as other species, have been dealing with the reality of microbial contamination since our first existence. Microbes such as *Escherichia coli* and protozoa like *Cryptosporidium* and *Giardia* are common in all wild animals and they are excreted into every stream. While wild animals have evolved to deal with these pathogens, humans do not fare so well.

Microbial Pollution

No discussion of microbial pollution would be complete without noting the work by John Snow, who is considered one of the founding fathers of modern epidemiology. During Snow's lifetime in the mid-1800s, cholera (the Black Death) was believed to be caused by chemical pollution and poor air; modern germ theory was not promulgated until the late 1880s. Around 1854, London had yet another outbreak of cholera. After careful investigation of the locations of outbreaks, Snow identified the Broad Street pump, one of many pumps delivering water from the Thames, as the culprit. After the handle to the pump was removed and other sources of water were secured, the epidemic was contained. This case is representative of several common themes used throughout this chapter: (1) location of your drinking water source is everything and (2) everyone lives downstream (from someone else's waste).

Surprisingly, not all bacteria found in warm-blooded animals are harmful to humans. One group of bacteria that receives significant attention is enteric bacteria, bacteria of the intestines that include the well-known

E. coli. Most varieties of *E. coli* are completely harmless to humans but are measured as "indicator organisms," since their presence indicates that fecal material and the other actually pathogenic bacteria that this material typically contains have entered the water.

"Natural" Pollution

While microorganisms are certainly important with respect to drinking water, it is also not uncommon for harmful, but natural, chemical agents to exist in some surface waters and certainly in groundwater. Depending on local geology, mineral deposits can contribute chemicals such as fluoride and arsenic to sources of drinking water, creating either short- or long-term health effects. A small amount of fluoride, less then one part per million, is good for our teeth, but excessive amounts can cause adverse consequences. But let us explore the example of arsenic pollution in more detail.

Arsenic pollution of drinking water obtained from groundwater sources is far too common a problem, yet it is completely natural. In 2007, one study estimated that approximately 137 million people in 70 countries drank from groundwater containing unsafe levels of arsenic, including the United States. The most serious case is the water supply in the Ganges Delta in West Bengal, India, and Bangladesh. Because of the need for a sanitary source of drinking water, millions of relatively deep groundwater wells were constructed, however, one in five of these wells tapped arsenic-contaminated water. Today, the contaminated wells have been identified and new wells use an installation-treatment procedure that largely removes arsenic from groundwater.

Industrial Pollution

While natural pollutants are certainly an issue, local or upstream industry can greatly complicate surface and groundwater pollution. Given our ingenious ways of making chemicals today, history has shown that health effects from some chemicals can take decades to document. Prior to the creation and implementation of the Environmental Protection Agency

(EPA) and the series of environmental laws that followed, there were very few regulations concerning what industry or anyone else could release into a waterway in the United States. Why is industry commonly located near or on rivers? The answer is that industry typically uses large volumes of water, and in the absence of water rights laws, this water is essentially free. The waterways also provide a very inexpensive means of shipping products to national and international markets. And, prior to the EPA and the Clean Water Act, waste disposal was relatively easy, since wastewater was diluted with river water—hence, the disturbing and incorrect expression, "dilution is the solution to pollution."

But in the old days, wastewater from industrial processes could either contain relatively nonharmful trace concentrations or highly toxic pure substances in high concentrations. Waste disposal greatly changed with the passage of the various forms of the Safe Drinking Water Act of 1974 and amendments in 1977, 1986, and 1996. While this act has several far-reaching dictates for cleaning up waterways, industrial pollution from wastewater disposal falls under the jurisdiction of the National Pollutant Discharge Elimination System (NPDES) of the Clean Water Act. The EPA defined which chemicals were controlled, and under the NPDES they strictly specified and monitored what could be released into a waterway. Permit violators could be fined, imprisoned, or have their contributing facility shut down. Implementation of the NPDES greatly changed how industry conducted business. In order to meet the requirements of the permit, many facilities were forced to install wastewater treatment facilities to remove pollution from the water prior to its being released into the environment. Some treatment systems were as simple as adjusting the pH of the water, while others required considerable effort and cost. In general, modern industry in the United States is doing well with wastewater treatment, but there are legacy polluted sites, such as Superfund sites (large-scale polluted sites under federal cleanup action such as Department of Energy sites at Hanford, Oak Ridge, and Savanna National Laboratories) created decades ago that may never be cleaned up. Given the necessity of locating industry near waterways, releases from these sites continue to pollute some rivers and streams.

Military and Governmental Pollution

Some of the worst pollution sites are still under remediation and come from a surprising source, our own government. I have worked on pollutant investigation and remediation projects at two U.S. Department of Energy national laboratories. While I probably know only the tip of the iceberg with respect to the problem, I have also heard reports from scientists in other countries. These problems arose during World War II and the Cold War. Generally, citizens of a country only hear after the fact about the environmental issues associated with government sites such as those operated by the U.S. Department of Energy (DOE) or the U.S. Department of Defense (DOD), or their equivalents in other countries. Local residents keep a close watch on the pollution and news of remediation efforts at these sites, but in order to understand how these environmental disasters were allowed to occur, we have to travel back in time.

Few of us alive today can truly appreciate the threat posed by Germany and Japan prior to and during World War II. The Allied governments put war-related industries in high gear in order to combat the threat of war and invasion. The well-known Manhattan Project spanned many secretive sites across the United States, fortunately often located in remote areas. Environmental concerns, such as long-term safe disposal of toxic or radioactive wastes were, understandably, of little to no concern at this point in history. Chemical wastes at these sites were buried in open pits or pumped into the ground. Because of the typical location of many sites near a river, these actions in years to come led to river pollution. Some of these sites were actively used as late as the 1980s. They are now our legacy. Many DOE and DOD sites across the nation are Superfund sites; other nations have similar contaminated sites.

One intelligent move that the U.S. government made at the time was to locate most of these sites in what were then relatively unpopulated areas of the country. As many investigations have shown, while this did not reduce the toxicity of the waste, it did lessen the risk to humans at the time, and often today as well. Currently, as our buried toxic wastes are slowly transported toward rivers by rain and groundwater, we face dangers to ecology and human health that may have no solution. Examples include rivers near the Oak Ridge National Laboratory in Tennessee and the Hanford Nuclear Waste Site in Washington State, which are being treated for all types of waste including radioactive, organic, and toxic metal waste. Time will record how many decades or even centuries it will take to remediate these sites. But to put things in perspective in the United States, a statement from Boris Yeltsin, the president of the Russian Federation from 1991 to 1999 summed up the former Soviet Union's pollution: "If Russia had American's pollution problems we [Russia] would have no pollution problems." In other words, although there are significant problems in the United States, the scale of pollution in other countries is far worse.

Defining Water Pollution

The moral of these stories about arsenic, industry, and government is that pollution comes in all varieties: natural and human-made, organic and inorganic, biological, and radioactive; and it comes from numerous sources: corporate industry, government installations, and even our kitchen sinks. Some cases of water contamination are relatively easy to clean up, while others require extensive treatment that adds to the cost of modern society. But in almost every case, modern industry can produce safe wastewater, clean up our rivers, and ensure the safe use of water downstream from pollution sources.

But what levels of water pollution do we need to be concerned with, since pollution seems to be everywhere? Actually, there is no natural source of perfectly "pure" water. Rainwater is generally considered the cleanest of our available sources of water but even it contains dust (and therefore dissolved minerals), dissolved gasses, measurable concentrations of DDT (dichlorodiphenyltrichloroethane) and PCBs (polychlorinated biphenyls) (carcinogenic chlorinated hydrocarbons used for decades prior to the 1970s), inorganic mercury, and many other chemicals that stay in the atmosphere for years or decades. Surface water is our next cleanest source, yet as we saw earlier, from a microbiology standpoint, untreated surface water is not exactly safe to drink. Scientists have expended considerable effort and cost to produce pure water in the lab, but oddly enough, this water does not taste very good, nor is it very good for you since drinking water should contain various electrolytes.

Since pure water is a relative misnomer, the key question is how "clean" does water have to be in order to be considered safe? Scientists measure the relative "cleanness" of water in terms of pollutant concentration, or how much pollutant is dissolved in water. A common unit most people can relate to is parts per million, meaning there is one part of pollutant per million parts of water. Illustrative examples of a part per million include 1 inch in 16 miles, 1 second in 11.5 days, and four drops of water in a 55-gallon barrel. Examples of an even smaller, but also important unit, as we will see in chapter 4, a part per billion include one penny in 10 million dollars, 1 second in 32 years, 1 foot in the distance of a trip to the moon, one blade of grass on a football field, and one drop of water in an Olympic-sized pool.

While 1 part per million seems like a small dose to worry about, water chemists and toxicologists actually consider it a very high unit of concentration. Many chemicals are in fact toxic at parts-per-million concentrations. In later chapters of this book, pollution problems look at parts per billion, parts per trillion, and even parts per quadrillion— levels at which we have great difficulty even measuring these toxic chemicals or naturally occurring necessary biological chemicals in our bodies.

Treating Water Pollution and the Lifetimes of Various Pollutants

So, clearly water pollution problems exist, some natural, some human-made. This brings up the question, "Do we treat our waste or just treat our drinking water?" Today, the answer in the United States and many countries is both. For centuries, we only had to choose to seek out a source of clean drinking water or to crudely treat it. Given our constantly increasing national and global populations, and the fact that we all live downstream, today we have to employ both practices to try to keep everyone's drinking water clean. In order to understand how this works, we first have to examine how contaminants behave in a river as water moves toward the ocean, the ultimate destination of all rivers.

Microbial Pollution

When a river or lake is contaminated by microorganisms, either from humans or from domestic or wild animals, nature immediately starts to clean up the contamination. Many of the organisms living in warm-blooded animals thrive while in the body, but do not survive well outside of it. Cooler temperatures in the water interfere with the metabolic processes and reproduction of these organisms, and natural predators kill a large number of them. Bacterial species such as *E. coli* survive longer than most human pathogenic bacteria, and these live a few minutes to hours outside of a warm-blooded animal. So, if fecal matter has entered the water, and if *E. coli* have perished, the pathogenic bacteria have mostly expired as well. The same cannot be said about viruses. Some viruses such as the human immunodeficiency virus (HIV) die almost instantaneously, while others (e.g., the hepatitis virus) can last for months to years in natural waters.

Chemical Pollution

Chemical pollutants are a completely different story. The most effective way to manage surface waters for chemical pollution is to treat the pollution prior to release into the river or lake. Thanks to the requirements based on the Safe Drinking Water Act and the Clean Water Act's NPDES permit system discussed earlier, U.S. industry, and the industry of many other countries, has become very effective at cleaning waste effluents, and industrial treatment processes are now well established and quite routine.

Many organic pollutants already present in surface waters can be biologically or chemically degraded by the water and dissolved oxygen. The main problem with decades-old pollution sites is that the once-waterborne pollutants have over time adhered to dirt particles and now reside in the sediment. If these pollutants remain undisturbed they are relatively safe, but if the sediments are resuspended, the attached pollutants can reenter the water and produce new toxic effects. One benefit of river systems over lakes is that rivers can self-clean by washing pollutants out to the ocean. This occurs every time there is a large pollutant spill in a river located anywhere near the ocean. This is not to say that "dilution is the solution to pollution," but when accidents do happen, the effect on local populations can be minimized by the rapid movement of the plume of pollution downstream. In lakes, on the other hand, years and even decades are required to replace the trapped water. Lakes, therefore, pose one of the greatest challenges to surface-water cleanup. This is especially true for the Great Lakes in the United States. The larger the lake, the longer the water turnover and resulting cleanup time. Lake Erie, the smallest of our Great Lakes, is the most polluted Great Lake in the United States.

Producing safe drinking water from a bacterial and viral standpoint is not a complicated process. Medical lore in Sanskrit, from circa 2000 B.C.E., states that "impure water should be purified by being boiled over a fire, or being heated in the sun, or by dipping a heated iron into it, or it may be purified by filtration through sand and coarse gravel and then allowed to cool" (the Sus'ruta Samhita and Ousruta Sanghita). It sounds incredible, but these 4,000-year-old practices are what many of our drinking-water treatment plants still use today. More recent advances in water treatment include the use of alum, a coagulant, as early as 1500 C.E. followed by the introduction of filtration in the 1700s and then sand filtration in the early 1800s. It is worthwhile to look at the processes used in modern drinking-water treatment plants and examine exactly how safe our drinking water is today.

Modern Water Treatment in the United States

We will limit our discussion to surface water, since drinking water sourced from groundwater is usually pathogen- and pollutant-free, with a few exceptions that will be discussed later. Location, location, location—to start with, the source of your drinking water is very important. The economics of water flow, i.e., the fact that water flows downhill dictates that drinking water treatment plants should be located uphill from the municipality to be served; this practice avoids pumping costs. It is also generally considered an unpleasant practice to locate the inlet of your drinking water plant at or below the outlet from your wastewater treatment facility. Everyone knows that natural water contains some turbidity, the suspended dirt that gives water a cloudy look. Surprisingly, the more dirt present, the easier it is to treat the water. In a treatment plant, first large floating items are removed by a metal screen; if leaves and other relatively large debris remain, the water is placed in a large basin and the debris is allowed to settle. Next, chlorine is added, usually in the form of sodium hypochlorite (NaClO), or common bleach. The reason for this is it is important to avoid creating a bioreactor in the treatment plant that actually grows more microbes than what originally entered the facility. The following step reveals why the presence of significantly suspended dirt is helpful. Alum [aluminum sulfate, $Al_2(SO_4)_3$] is added to the water and slowly mixed. The aluminum forms a very sticky chemical [aluminum hydroxide, $Al(OH)_3$] that adsorbs to suspended dirt particles. This forms a mat that filters down through the water column and removes other chemicals, dirt, and very small particles. At this point, the water appears clean, but the EPA requires it to be a bit clearer. In addition, bacteria, including pathogens, have yet to be completely treated.

The water passes next through a sand filter, the same sort of sand filter mentioned in the Sanskrit tradition from 2000 B.C.E. This is the key, required feature in almost all drinking-water treatment plants in the United States. Based on size, sand filters remove any remaining "floc" or small particles from the aluminum treatment, as well as microbes and viruses that remain in the water. Finally, a last chemical treatment is undertaken as chlorine or ozone is added to kill any remaining microbes. Note how very simple this treatment process is. The water produced in this very economical system is high quality and meets all of the EPA's water treatment goals: (1) the removal of pathogens, (2) the removal of turbidity or dirt in order to give an aesthetically pleasing appearance and remove particles where microorganisms can hide, and (3) the removal of other pollutants, usually in the settling basin but sometimes with additional treatment. To ensure adequate treatment, the quality of the water and the concentrations of chemicals added to it, or naturally present, are constantly or regularly monitored, depending on the chemical. This process is so successful that rarely has there been an outbreak of disease that can be traced to an improperly functioning treatment plant in the United States.

One interesting difference between European and U.S. drinking water treatment plants is the disinfectant used. Since the 1970s, Europe has generally been opposed to the use of chlorine and has instead relied on ozone to disinfect drinking water. (In fact, Clorox® is generally not even available in Europe.) Ozone is a very reactive form of oxygen in which three oxygen atoms are "uncomfortably" bound together; it almost instantaneously kills any microbe, virus, and microbial cyst upon contact. This is effective, but because ozone is so reactive, no residual disinfecting agent that was added remains as it passes through the city distribution system and into the tap in your kitchen. A health risk of using ozone can therefore result if your distribution system is compromised by a crack in the water pipe that allows microbes to enter the distribution system. With no residual disinfectant, the outbreak of a microbial disease by drinking this water is much more likely.

On the other hand, chlorine maintains disinfecting power in drinking water for a longer duration, as evidenced by the chlorine taste in some city waters. Chlorine can be added in two ways: (1) as chlorine gas (Cl_2), which reacts with water to form hypochlorite (OCl^-), a reasonably good disinfectant, or (2) by adding sodium hypochlorite ($NaOCl$) to water. Hypochlorite remains in water for a considerable amount of time. When a very small amount of ammonium is present, hypochlorite forms even more stable and effective disinfectants.

Chlorine-based chemicals are not, however, as effective a disinfectant as ozone. Both ozone and chlorine are very effective at killing living bacteria and viruses, but dormant bacteria are more resistant. For example, ozone can kill *Cryptosporidium* cysts but chlorine is less effective at doing so. Since the 1990s, a few *Cryptosporidium* outbreaks have occurred in chlorine-based treatment plants in Canada, England, and the United States, which is why plants are being encouraged, and sometimes forced, to use ozone if their source of water is ever found to contain these cysts.

But there is a more fundamental dilemma with chlorine treatment. All surface waters contain some natural organic matter, the molecules from dead plants and animals. There is absolutely nothing harmful about having these unavoidable organic molecules present, but an issue can arise when chlorine is added. Many disinfecting forms of chlorine react with natural organic matter to form a class of compounds known as trihalomethanes, which are highly chlorinated organic compounds. History has shown, and Europe has duly noted, that organic compounds containing chlorine frequently pose a threat to animal and human health. Many are in fact carcinogens. Hence, chlorine treatment is also potentially prob-lematic. To summarize, ozone provides the most thorough, primary dis-infection but does not disinfect all the way to your tap. Chlorine, on the other hand, typically provides adequate disinfection that continues all the way into the glass of water in your home, but it can create carcino-gens in drinking water. So, which is best?

The answer to how to treat or even whether you should treat your drinking water results from undertaking a risk assessment of treatment options. These risk calculations clearly show that both ozone- and chlorine-treated water are better than untreated drinking water. If you do drink untreated water, statistically you may eventually contract a fa-tal disease, especially if you live in a highly populated area, as many of the world's people do. While chlorine treatment could in theory cause cancer, well-documented, long-term studies have shown that the risk of contracting cancer from the parts-per-billion organochlorine chemicals present in the water is very low to nonexistent even after drinking the treated water for 70 years. This is certainly less than your chances of contracting a deadly disease from drinking untreated water, and as it turns out, the chances of dying from many other causes is much greater than the risk of cancer caused by carcinogenic drinking water. In light of all the pros and cons of ozone versus chlorine, Europe continues to pri-marily treat with ozone but is now starting to add a trace of chlorine for a disinfectant residual in the water distribution system. Likewise, the EPA now prefers the major microbial treatment to be done with ozone, fol-lowed by a residual of chlorine. In any event, drinking-water treatment plants have been primarily responsible for reducing deadly disease out-breaks in Global North countries, and Global South countries are making great strides in building treatment plants to reduce the burden of disease.

The inexpensive cost of modern drinking-water treatment is also wor-thy of mention. In Global North countries we have excellent access to clean and safe drinking water, but at what cost? In the United States, treat-ment plants are paid for by state and federal grants (i.e., your tax dollars) and by municipal bonds that you help pay off in the form of monthly water and sanitation fees. The actual cost of a drinking-water treatment plant for a city population of 100,000 residents varies depending on the quality of the raw water available but averages around $20 million to $30 million dollars. This cost is somewhat buffered by spreading the expense over time and across the population, meaning that each household pays only a few dollars a month for its water service. The household cost of a gallon of clean drinking water ranges from less than a penny to a few pennies depending on the source of water and necessary cleaning technology, while the cost to treat a gallon of sewage wastewater is several pennies, again depending on the technology used. If you are wondering about the aggregate costs where you live, these numbers are fairly linearly scalable. Regardless of location, overall the cost remains very low.

Compare this to the cost of the trendy bottled drinking water in the grocery store. A quick survey of my grocery store's personal-sized bottled water yields a range of about $2.50 to $7.16 per gallon, while a similar volume of tap water can cost less than one cent. Consumers therefore generally pay between 240 and 10,000 times *more* per unit volume for bottled water than for tap water. If you are in the secure area of a large airport, prices increase exponentially to $30 to $40 a gallon. And we complain about the price of gasoline! Not surprisingly, a study completed in 2006 estimated the global market for bottled water at $60 billion, with a volume of 3.0×10^{10} U.S. gallons, produced by approximately 800 to 1,000 companies. Ironically, the country with the safest municipal drinking water, the United States, is the largest consumer of bottled water, with the average American consuming 21 U.S. gallons a year. But why pay these exorbitant prices for water when there is a chilled water fountain near every bathroom?

Clearly, bottled drinking water is a booming business. There are some interesting claims as to the supposed benefits of different brands, but given that approximately 10 to 15% of the cost of the bottle in your hand is due to advertising, the claims are not surprising. There are plenty of concerns that have been raised about bottled water, particularly about the energy used to produce the plastic and transport the bottled water, not to mention the chemicals that may leach from the plastic bottle.

As it turns out, municipal drinking water systems have received a bad rap in recent years due to the very rare presence of *Cryptosporidium* spores in treated water, as discussed above. This rare occurrence has been hyped up by some groups and has led to rapid sales of bottled water. Some bottled water, like the multigallon source in your grocery store is local drinking water with additional treatment, while others range from local spring and/or mineral water to water imported from around the world. Bottled water can have vitamins and other additives, and some companies even market completely deionized water. The consumption of bottled water in the United States quadrupled between 1990 and 2005. Interestingly in the United States, the Food and Drug Administration (FDA) regulates bottled water while the EPA regulates the quality of tap water through the Safe Drinking Water Act.

So which is purer and safer? It depends on the water source and extent of treatment. Bottled spring and mineral water are inspected, but in 1999, a controversial National Resources Defense Council study found that several sources have been shown to contain natural arsenic, other synthetic organic contaminants, and bacteria levels above those deemed safe by the EPA. Excessive fluoride is not uncommon in groundwater and is difficult to remove. In the early 1990s, the carcinogen benzene was found as a contaminant in Perrier mineral water. In addition, there are known instances of organic chemicals, including endocrine disruptors, leaching out of some plastics used in water bottles. And of course, we have not even begun to discuss the waste of petroleum hydrocarbons to make the billions of plastic bottles that end up in our recycle bins or, worse, landfills. Numerous bottles also wind up in rivers that carry them into our oceans, where they contribute to the growing Pacific and Atlantic garbage patches; in the Pacific the surface area of the patch is approximately the size of Texas. But the waste of plastic resources mentioned here is a bit of a distraction from our success story: the availability of clean and safe water, without the need for bottled water. We'll return to the subject of bottled water in chapter 4.

Water from municipal treatment plants is not perfect. Some of its greatest weaknesses stem from infrastructure issues, particularly broken water delivery lines that lead to contamination and soldered pipe connections in your own home that leach lead (a toxic metal). One simple and very inexpensive solution to the potential dilemmas of attaining clean domestic water is installing a commercially available water filter on your kitchen tap or in your refrigerator that will still deliver safe drinking water for a fraction of a penny. But this is not to say that bottled drinking water does not have its place, it should just be very limited. Bottled drinking water can be a daily necessity in certain parts of the Global South, where large-scale municipal drinking water systems are not currently available.

Worldwide efforts have made bottled water reasonably affordable to citizens in these countries, far less expensive than in the United States and Europe, where expenditures are instead put toward maintaining treatment plants. The health of citizens in many countries depends on bottled water, and this will continue until permanent, large-scale drinking-water treatment plants, along with distribution systems, are built.

Status of Global Access to Safe Drinking Water

Global North countries around the world have made great strides in providing safe drinking water to their citizens, as well as cleaning up the sources of their water, rivers and lakes, that will be discussed in chapter 2. U.S. waterways in particular are no longer considered open sewers and do not catch fire as occurred several times in the 1960s on the Cuyahoga River in Cleveland, Ohio. Our goal now is to extend our success to the Global South.

Many national governments, international organizations, and informed citizens throughout the world, consider safe drinking water to be a basic human right. Numerous campaigns are well underway around the world to fulfill this goal. In doing so, we will most likely follow the approach that many Global North countries have used. First, one must attempt to obtain drinking water from a pristine source, most likely groundwater that is typically naturally free from pathogens and chemical contamination because of its isolation from surface sources, with the exceptions noted previously. When this first, preferred approach is not possible, surface water should be treated to meet safe drinking standards, knowing that pollution sources exist upstream. A necessary and complementary component is to remove waste sources from upstream which will be the subject of chapter 2. This is the main long-term objective of assisting the world's less privileged populations today. But the scale of dire need is staggering.

Here are the shocking statistics. There are approximately one billion people without access to safe drinking water, defined as a water availability of 20 liters (approximately 5 gallons) per person per day at a distance of no more than 1,000 meters from one's home. It is estimated that every day around 5,000 to 6,000 children die from waterborne diarrheal diseases such as dysentery and cholera. That is two million instances of child mortality each year, or 250 every hour, that could be prevented. While those of us in the Global North pay only pennies, or less, for a gallon of safe drinking water, citizens in Global South countries can pay ten to hundreds of times more. Even if the funding was available today, it would take years to provide all residents of Earth with safe drinking water. We are, however, working diligently on the problem that will likely define the twenty-first century. In the short term, aside from relatively costly bottled water, many companies have developed personal purification devices such as the relatively inexpensive LifeStraw, the Hydro-Pack water filter pouch, and Pur packets that can be used to supply small volumes of safe drinking water. One of the simplest and most common, although not very pleasant, treatment methods is to add a couple of drops of Clorox® to each gallon of water and allow it to sit for a short time. These devices and methods mostly treat microbial contamination, but do nothing for natural or industrial chemical contamination.

Also related to safe drinking water, are the 2.5 billion to 3.5 billion people without access to basic sanitation (and these people are not just out camping, as an engineering friend of mine notes). That's 4 to 6 of every 10 people on Earth. Remember, we all live downstream! As we will see in chapter 2, there is an inseparable relationship between safe drinking water and sewage treatment, since widespread lack of effective waste disposal contributes to the diseases contracted from unsafe drinking water.

So far, I have described the challenges associated with providing safe drinking water, but what about the scarcity of water in general? Climate change is affecting some of the most vulnerable regions of the planet. So, not only do we need to worry about providing safe drinking water, we also need to focus on water conservation practices, and not only in less fortunate parts of the world. Look at news reports from 2014 (Ayres 2014) concerning the long-term drought in the North American southwest and the resulting tenuous allocation of scarce water.

To put this in perspective, consider the amount of water it takes to produce the following commodities. One kilogram of coffee requires 20,000 liters (5,000 gallons) of water. One quarter-pounder hamburger requires 11,000 liters (2,800 gallons) of water. One kilogram (one-half pound) of cheese requires 5,000 liters (1,300 gallons) of water. Drought promotes eating lower on the food chain, since 1 kilogram (one-half pound) of rice requires only 5,000 liters (1,300 gallons) of water and 1 kilogram of wheat requires only 1,000 liters (250 gallons) of water. Water scarcity is even listed as a future threat to our national security. But this is a topic for other authors and books. A clean drinking supply is a much more surmountable goal.

So, what will it cost to provide the world with safe drinking water and adequate sanitation? Given the scale of the problem, numbers vary greatly. The widespread, global recognition of this goal is reflected in the United Nations (UN) Millennium Development Goals and subsequent, related documents. The relevant Millennium goals are as follows (World Bank 2005):

Goal 4: Reduce child mortality by two thirds between 1990 and 2015.
Goal 7.c: By 2015, halve the proportion of the population without access to safe drinking water and basic sanitation.

These relatively simple statements do not, however, demonstrate the enormity of the challenge ahead or the range of estimated costs. Numbers vary considerably among sources, but it is clear that several issues have to be addressed simultaneously. Efforts to improve education, health, and the environment (water and sewage treatment) must be concerted. Cost estimates per year range from $10 billion to $30 billion for education, from $20 billion to $25 billion for health improvements, and from $5 billion to $21 billion for the environment. In total, the annual funding range to achieve safe drinking water and sanitation, and thereby greatly reducing waterborne diseases, falls between $35 billion and $70 billion per year through 2025, which is the timeframe by which the UN aims to achieve this goal. Most of us cannot relate to these large numbers, but governmental budgets can add considerable perspective.

One measure of a government's or a country's wealth is its annual gross domestic product (GDP), the market value of all officially recognized goods and services produced within a country in a given year. Table 1.1 gives the annual GDP for the top four economies in billions of U.S. dollars. (Note that the 27 member nations of the European Union are aggregated into one economy.) In contrast to GDP, each government's budget, still in the billions of dollars, is substantially less. Citizens of many countries love to first criticize budgetary items such as the military and

TABLE 1.1 Gross Domestic Products for the Top GDP Countries

Top GDP countries	GDP in billions of U.S. dollars in 2011	Government's total budget in billions of U.S. dollars	Government's military budget in billions of U.S. dollars	Military percent of total government budget	Government's foreign aid to other countries in billions of U.S. dollars	Foreign aid percent of total government budget
EU 27 member states	17,600	3830	−260	6.8	73.6	1.9
United States	15,100	3830	655	17.4	49.5	1.3
China	7,320	828	92	11.1	?	?
Japan	5,500	1100	56	5.1	7.04	0.64

? Questionable data or no reliable data are available.

foreign aid. These are also given in the table, with a column beside each showing their respective percentages of a given country's total budget. A few numbers stick out. First, the United States spends considerably more of our tax dollars on military expenditures but ranks only second in foreign aid. One could imagine the positive effect of taking some of the military budget and spending it on foreign infrastructure such as water and wastewater treatment, instead of securing oil for our excessive automotive use and low-mileage automobiles.

How many regional conflicts and actual wars would this prevent? And remember that the CIA considers water security (scarcity and quality) a risk to U.S. national security! If the cost of the UN Millennium Development Goals were to be spread proportionally among the top nations in the world, with only minor rearrangements of how these countries currently spend their money and small increases in foreign aid, the entire world could easily and relatively quickly have safe drinking water and sanitation.

But it is always easy to pick on government budgets, and in the end, government expenditures are funded by citizen taxes. What are you willing to give up to provide others with safe drinking water? The following are some options that everyone could choose to give up in order to donate the resulting savings to international efforts:

- U.S. citizens spent $21.7 billion on bottled water in 2011 when the water in their kitchen or the closest water fountain is perfectly safe and far less expensive.
- The average U.S. citizen consumes 3.1 cups of coffee each day that translates to $164 billion per year. The total annual sales of specialty coffee in the United States is $18 billion. If everyone cuts back one cup a day and donates these savings, it would easily account for billions of dollars.
- U.S. citizens spent $184 billion on fast food in 300,000 restaurants in 2010. If everyone reduced this amount by 10%, it would yield almost $20 billion that could go to Global South nations.
- Annual alcohol sales in the United States in 2010 were in the neighborhood of $400 billion.

My point in listing these examples is that most citizens in the Global North live a very good life. If everyone cuts back a little and donates the savings, we can easily supply safe drinking water and sanitation to all residents of planet Earth.

So, where should the money go? First, contact your local congressional representatives and demand foreign aid for water and sanitation, and specify that military aid should be secondary to these fundamental rights. If you are looking to donate for an international agency, the list given on the UN Millennium Development Goals website (UN Millennium Development Goals 2014) is a great place to start.

But I need to add one final word on building water and wastewater plants in the Global South. In the recent past, there have been disagreements about efforts to privatize these facilities and water-distribution networks. Privatization comes in three forms: management contracts, lease contracts, and concessions. Management and lease contracts, where professional engineering runs a city's services for a fee, are becoming common in the Global North and are a must in Global South countries. Concession of water rights is where a population gives the outright ownership of the source of the water, the water itself, and its distribution system to a private company for hire to supply water at a certain cost to the affected population. Full concession of water and facility rights and ownership has experienced mixed reviews because of excessive pricing, poor maintenance of the systems, and quality of the water. Cities in

Global South countries and the UN programs that have had grave problems with this approach are currently, and understandably, trying to avoid this type of arrangement. There are many highly qualified engineers and companies waiting for international investment to build water and sanitation facilities in Global South countries, and all we need is the funding. We are at the cusp of giving inexpensive, clean, and safe drinking water and wastewater treatment to the world. The UN Millennium Development Goals, which are being replaced by the Sustainable Development Goals in 2015 (UN Sustainable Development Goals 2016), have a target date of 2025, and prospects for meeting this accomplishment dare are looking good.

TWO

Effective Treatment of Our Wastewaters

COLOGNE

In Köhln, a town of monks and bones,
And pavements fang'd with murderous stones
And rags, and hags, and hideous wenches;
I counted two and seventy stenches,
All well defined, and several stinks!
Ye Nymphs that reign o'er sewers and sinks,
The river Rhine, it is well known,
Doth wash your city of Cologne;
But tell me, Nymphs, what power divine
Shall henceforth wash the river Rhine?

—SAMUEL TAYLOR COLERIDGE

ON MY JOYFUL DEPARTURE FROM THE SAME CITY

As I am rhymer,
And now at least a merry one,
Mr. Mum's Rudesheimer
And the church of St. Geryon
Are the two things alone
That deserve to be known
In the body and soul-stinking town of Cologne.

—SAMUEL TAYLOR COLERIDGE

The above poems put civilization's need to develop wastewater systems in a clear and pungent perspective. Chapter 1's discussion makes obvious the inseparable link between clean and safe drinking water and effective wastewater treatment. Every country that I know of has followed the same path through their development; first, clean your drinking water; then, when the population reaches a critical number, clean your wastewater. Our ability to effectively handle our waste is a surprisingly recent development. Civilizations, as defined by anthropologists, first came about circa 5000 B.C.E.; and many civilizations have been studied, and defined, by their garbage dumps. How will civilizations of our age be defined, by our pollution or our remediation?

Prior to the development of any concentrated population that constituted a civilization, all rivers were relatively clean. Yes, animals excreted into the water but the amount of waste, organic and bacterial, entering the stream did not exceed the stream's ability to handle the waste—the waste did not interfere with the normal life of other animals in the stream. As we noted in chapter 1, rivers have built-in cleansing power; they wash their contents downstream, where it is diluted with more and more water, pathogens die, and natural organic waste is oxidized. If a chemical of interest degrades, the waste concentration becomes even more dilute and can disappear completely. Rivers were, and are, very adept at handling small volumes of natural organic waste.

As this chapter is the longest in the book, it needs an introductory outline or roadmap. I'll start with a little chemistry of dissolved oxygen in water and how it relates to organic waste in the water. Then, I will describe the history of our second success story, treating our domestic wastewater prior to releasing it to a stream. Next, I will discuss possible solutions for the future of wastewater treatment, and finally, I will provide examples for countries that currently do not yet have the resources to build and maintain high-tech sewage treatment facilities.

Dissolved Oxygen

The defining chemical that regulates how fast natural organic waste, specifically sewage, will degrade in a body of water is the amount of oxygen in the water. Diatomic oxygen is very abundant in the atmosphere. Depending on humidity, it comprises about 19 to 20% of the atmosphere, which translates into approximately 800 parts per million using the Ideal Gas Law from chemistry and not the 20,000 parts per million that one would initially expect. The atmospheric concentration of oxygen has been relatively constant for several hundred million years; this has been key to giving more advanced forms of life time to evolve. Aquatic forms of life evolved before and alongside terrestrial organisms; however, life in the aquatic world was and is not as easy. Because of chemical differences between water (H_2O is polar) and diatomic oxygen (O_2 is nonpolar), very little oxygen is actually dissolved in water: depending on water temperature, only around 8 to 10 parts per million or milligrams per liter. Comparing the approximately 800 parts per million available to life forms in terrestrial systems to the 10 parts per million accessible to species living in aquatic environments clearly shows why air breathers have been able to evolve to perform more advanced activities, including leisure time to read books like this one. More important is the fact that a much lower concentration of oxygen is available to degrade any waste added to a stream. There are plenty of aquatic microbes that can consume organic compounds, but oxygen must be present for these organisms to perform this activity.

The amount of dissolved oxygen in a stream is a straightforward, simple, chemical measurement, but we need a way to relate this to the dissolved oxygen needed to oxidize biological waste. This parameter is called the biochemical oxygen demand (BOD) and refers to the total amount of dissolved oxygen, in milligrams per liter or parts per million, that is needed by microbes to oxidize the waste. Here's the problem. We can create far more BOD in a stream than there is dissolved oxygen. No matter what we do, a water body can contain, at most, 8 to 10 mg/L of dissolved oxygen. This is not a problem in natural, unpolluted streams, since they usually contain less than the same amount of organic matter to be degraded. But typical waste from untreated domestic sewers contains around 150 mg/L—roughly 15 times more than the dissolved oxygen present. Industrial waste is even more problematic, since its BOD levels can reach hundreds to thousands of parts per million of dissolved oxygen. When this happens, the stream ecosystem essentially dies because oxygen-dependent animals are unable to survive as microbes take all of the dissolved oxygen out of the water. Streams will take in more and more oxygen from the atmosphere to restore equilibrium as dissolved oxygen levels in the water fall, but the re-aeration process typically cannot keep up with the rate of microbial consumption of oxygen. The area of a low- or no-oxygen zone can extend for miles to hundreds of miles, depending on various factors. Large rivers can take on considerable amounts of raw sewer water with little effect to the ecosystem, but the human disease potential contained in these waters is still very high. Smaller streams can be destroyed even by very minimal waste inputs. A long distance downstream, or given extended recovery time, the stream can eventually recover.

Historical Efforts to Deal with Wastewater

In all cases of remediation, whatever the system or the pollutant, the first thing that has to be accomplished is removal of the source of pollution. How can you possibly treat or clean up a system that is still receiving inputs of pollution? Remediation in the case of wastewater, or in other words, putting a stop to the dumping of raw sewage into our rivers, has been a relatively recent development. As noted in the chapter 1, humans have had to address their sewage waste since the first human settlement, no matter how temporary. Far back in our history, nomadic people simply packed up and moved to a new place. But as civilization grew and we collected more and more possessions, relocating was not as easy; thus, our first environmental problem was created—a need to deal with our sewage waste.

Depending on the location in the world, there is evidence of drinking-water distribution, storm water, and sewage removal systems in very large and reasonably developed cultures going back as far as 4500 to 2000 B.C.E. While these systems were fairly well developed, there was almost always a link between storm water drainage and sewage removal, a connection that still remains problematic today. Depending on location, topography, and water access, water and sewer removal developed at different times and at different rates throughout the world. In Western civilization, many large populations did in fact develop very effective systems for conveying water and waste, including the Mesopotamian Empire, the Indus and Aegean (Isle of Crete) civilizations, Egypt, and Palestine. As the Roman Empire grew, and similar developments sprang up in Greece and China, so did civil services. As noted by many historians, however, when the Roman Empire fell, so did the concepts of bathing, basic sanitation, and engineered water and sewage systems.

As Europe slowly redeveloped after the Dark Ages, during which disease and death were commonplace, an odd mixture of sanitation methods evolved. They centered on a storm-water-removal system, basically consisting of the middle section of streets and cesspools although in a few cases underground piping was used. If a home did not have the "luxury" of having a cesspool under their dining room floor, they used the street to dispose of night pots and all other waste. Everyone has probably seen a movie in which a knight or musketeer laid down his cape for a lady to cross the street. That was not mud they were covering up, and they probably did not promptly put the cape back on.

All forms of waste were disposed of in open-street drainage systems and were periodically washed out by rainwater. Of course this material ended up in the nearest stream, where downstream someone was obtaining drinking water. Without proper sanitation, people were forced to "relieve" themselves when and where the need arose, so much to the point that etiquette books of the time (circa 1500s to 1700s) contained statements such as *It is impolite to greet someone who is urinating or defecating*, and *If you see someone relieving themselves, you should act as if you had not seen them at all*. This was not a good time to be alive, nor was Europe the tourist destination that it is today.

For families that did have access to cesspools, they usually relied on an open hole or pit below the floor of each home. By the early eighteenth century, almost every European home had a cesspit beneath the floor. Depending on the geographic location, some pits were lined to prevent percolation into the groundwater while others were drained manually. More than a few floors covering the cesspit gave way during the dinner hour, certainly spoiling the evening for that family. There are even records of drownings—yikes! Sewer water and solids from the many cesspools quickly found uses as irrigation and fertilizer for farming. Disease outbreaks were still rampant across Europe, but amazingly it was not attributed to poor sanitation. The air was foul in all of Europe, not only from the ever-present raw sewage, but also from the widespread use of coal as a fuel. Nowhere was as bad as London.

Remember that disease theory had yet to be promulgated. The prevailing wisdom was instead that night air caused many diseases and death. In most Western cultures, people believed that night vapors called "miasma" rose from the soil and spread disease. Therefore, during the mid- to late 1800s residents of Europe closed their doors and windows to keep out miasma. In doing so, they locked in the vapors from their cesspool. Deaths from diseases such as cholera were common, most notably the outbreaks in London in 1854 (as mentioned in chapter 1). There were also numerous cholera outbreaks in other parts of the world with similar causes: in Bengal in 1816, in Russia from 1829 to 1851, in Russia from 1852 to 1860, in Europe and Africa from 1863 to 1875, in Europe and the Americas from 1881 to 1896, in western Europe from 1899 to 1923, and in Indonesia from 1961 to 1975. Between 1991 and 2009 there were again notable and numerous cholera outbreaks in parts of the world with inadequate sanitation.

Research in the mid-nineteenth century helped shed light on the true cause of many disease outbreaks. The work of John Snow attributed the cholera epidemic of 1854 to the water from one problematic well in London. Then, between 1860 and 1864, the well-designed experiments of Louis Pasteur led to the development of germ theory. Subsequently, Robert Koch's postulates demonstrated a microbial cause of disease. Together, these discoveries turned the understanding of the source of diseases like cholera away from bad night air to germs or some specific agent present in the water. The resulting connection to poor sanitation did not require much time for government and citizens to accept and they began to ask, could the causative agent be something in the water that they could not see?

Post-Medieval Waste Disposal Methods in Europe

Besides high population density, oddly enough, one of the leading factors in the increase of London's, and later Europe's and the Americas', poor sanitation problems was the invention and advancement of the water closet and toilet. When human waste was collected in household cesspools, it was compact and more solid than liquid and therefore easy to clean out and transport to local farm fields for use as fertilizer. The toilet and water closet added as much as 10 times the weight in water during flushing and thereby created an additional waste volume management problem.

But, let us start from the beginning. There have been sitting toilets as far back as 2100 B.C.E. in Egypt, while the mechanized versions are more recent. England was the first to enter the modern age of the semi-flush toilet in 1596, when Sir John Harrington, godson of Queen Elizabeth I, invented a raised cistern that had a valve for flowing water. Queen Elizabeth promptly installed this invention in her palace at Richmond and had a cover of velvet and lace placed on it. The price was 6 shillings 8 pence. Subsequent, notable modifications included those by Alexander Cummings in 1775, which included a better S-shaped disposal system to suppress odors, and by S. S. Helior in 1870, which resulted in the true flush toilet. After these modifications, the emphasis was placed more on aesthetics and decoration such that toilets became known as "soup bowls." But the increased volume from the use of flushing water overtaxed the capacity of household cesspools and they eventually overflowed into the streets. The British Sanitary Act of 1845 required all new dwellings to have water closet drains connected to street sewers. The following 3 years saw the construction of 146 miles of sewer system in Liverpool that directed waste to the River Mersey, which was, unfortunately, the source of drinking water. The large volume of waste carried by this sewer system and upstream sewer inputs lead to the Great Stink of 1858, when the stench was so bad that parliament could not meet. After about 15 years of construction, the main drainage system of London was redirected to the south side of the Thames and released at high tide to aid in the natural transport of sewage out to sea. Recall that several cholera outbreaks had occurred by this time.

Early American Waste Disposal Methods

To backstep a moment in our history lesson, Americans were slightly behind Europeans in identifying disease-causing sanitation problems and in developing proper sanitation. The population and urban densities were lower in the Americas, and cesspools were not as common. By the mid-1850s, sewage-handling systems consisted of open ditches dug in the middle of unpaved streets that received all types of waste. The developed eastern portions of the United States and Canada were subject to heavy rains that overflowed the common street sewers and flushed their contents back into homes and gardens. It was as late as the 1870s before places such as Brooklyn had adequate sewer systems completed. Boston's first sewer was constructed in 1876. These sewage-handling systems, like those in London and Europe, had no treatment facilities; they only provided disposal in the sea or harbor. Needless to say, the seafood in these areas was highly contaminated.

The Evolution of Modern Sewage Treatment

The timeline for the invention, development, and implementation of sewage treatment is not linear and depended upon a place's population density, proximity to flowing water and the sea, and many other variables. In all cases, the driving force was fighting outbreaks of cholera, the scourge of the 1800s. In some areas, reasonably effective sewage handling was implemented as early as 1860 with Jean-Louis Mouras's invention of the septic tank. Another technological innovation we will discuss later is the invention of the trickling filter by Edward Frankland as early as 1868 and used very effectively for at least a century. An experimental facility in Massachusetts created and installed the first sand filter in 1887. So while some progressive places had reasonably effective sewage treatment, others had none.

The development of sewage treatment was an end-of-the-pipe technology, meaning the treatment plant was placed at the end of the sewer drainage system. The first systems were based primarily on the biggest bang for the buck (the biggest return for the cost). One of the easiest ways to remove half of the biological waste produced by homes, or in other words the biochemical oxygen demand (BOD), was simply to place the wastewater in a holding tank and let the solids settle out. This was called primary treatment and was the basis of the earliest attempts at sewage treatment. In much of the United States and Europe, this treatment was used by many municipalities as early as the early 1900s through the 1960s, but surprisingly, even today this is still the only treatment method some locations have. One of the more recent ones is discussed below.

But what about the remaining BOD and the microbes associated with it? High amounts of chlorine, in the form of chlorine gas or chlorine bleach, used to be added to kill most of the microbes and then the wastewater was released to a receiving stream. These actions helped prevent disease but did little to maintain the ecology of the stream. Remember that typical domestic BOD levels are around 150 parts per million while streams have only about 8 to 10 parts per million of dissolved oxygen. The deficit of at least 140 parts per million BOD will destroy a stream and the aquatic wildlife it contains for many miles downstream.

The settling tanks mentioned above operate by a physical mechanism: the more dense BOD particles settle out of the water and move to the bottom of the tank while the cleaner water flows out of the top of the tank. Since this settling process occurs naturally, this is a very easy and inexpensive process. The removal of dissolved BOD, however, is far more of a challenge and requires biological treatment.

As mentioned above, the first biological treatment for settling tanks was the trickling filter treatment invented in 1868. Slow adoption of this type of treatment was a result of the lack of laws requiring their installation, their added cost, and the greater operating complexity they created. In developing these systems, engineers simply capitalized on the same microbes that degrade BOD in streams but introduced them into an engineered system. Trickling filter systems operate by slowly pouring water over a 5- to 10-foot-deep layer of rocks or rock-shaped media held in a large tank. As BOD-containing water is passed over the rocks, microbial colonies grow by converting the usable, nutrients dissolved in the water into cell masses. Wastewater exiting the trickling filter has lost most of its BOD. Thus, we have transformed dissolved BOD into solid cell masses. As the cell masses grow, microbial cell mats fall (sluff) off of the rocks and flow out of the system with the water. Hence, we need another settling tank, very similar to the one initially used to remove large particulate matter or solid BOD, to remove the microbial cell mats. The process to this point can remove approximately 90 to 95% of the BOD: 45 to 50% as solid BOD in the first settling tank and 40 to 45% as dissolved BOD through the trickling filter.

You may recall from chapter 1 that most pathogenic organisms do not live long out of warm-blooded animals; by retaining the water in these paired settling tanks, many of the pathogens die. To ensure the clarity of the water and the removal of pathogens, after the second settling tank, the fairly clean water is passed through a sand filter. This is in fact the same type of sand filter mentioned in the Sanskrit notes from 2000 B.C.E. quoted in chapter 1, which was reinvented in 1887, as mentioned above. Today, as a final touch, chlorine is added to ensure complete sanitation. Then, after adequate contact time to ensure sanitation, excess chlorine is removed and the now treated water is released to a nearby stream or farming operation for irrigation. More modern treatment plants avoid chlorine and instead use ultraviolet lamps to irradiate and thereby inactivate microorganisms.

Varying by country, these treatment options appeared at different times between the late 1800s and today. In the United States, these treatment processes have been regulated by the Clean Water Act and the contained NPDES (National Pollutant Discharge Elimination System) permits since the 1970s. These treatment practices have been highly effective, in the United States in particular, in preventing disease and cleaning up our rivers. All is not perfect in Global North countries, however. For example, the Canadian city of Victoria, B.C., in this decade made a highly embarrassing admission that for years it has dumped 34 million gallons of raw sewage each day into the Strait of Juan de Fuca. I should note that in 2009 the city agreed to build four treatment plants, at a cost of $1.2 billion to solve this problem. But it seems that even modern cities that had been thought of as "green" have skeletons in their closets.

Biological Treatment Options

The biggest change in the type of treatment plant described above has been the adoption of a different type of biological treatment. In 1913, two U.K. engineers developed a process known as activated sludge, in which an ecosystem similar to that in a stream is created in a large highly mixed tank, with only the addition of dissolved oxygen. These systems were, however, not commonly adopted to replace trickling filters until the 1970s to 1980s. Current activated-sludge processes reflect nuanced development over 80 to 90 years, and they are considered state of the art for sewage treatment in the Global North. Compared to trickling filter technology, activated sludge systems are more stable and controllable and offer better treatment of domestic sewage wastes, but they require greater costs and technical knowledge of the system.

A number of municipalities have sewage treatment systems that also receive nontoxic food-processing waste. These types of wastes are even more degradable by the microbes found in sewage treatment systems, since they have not even been digested by humans and have far more energy value and require more dissolved oxygen to degrade them;

therefore, they can be far more dangerous to stream chemistry and ecology. For example, the BOD of incoming domestic sewage flow to a treatment plant is around 150 to 200 mg/L. Food-processing waste can have a BOD as high as 10,000 mg/L, and as a result is even more damaging to streams. Yet, both types of waste can be effectively treated in the same treatment plant.

While these modern systems are very effective at treating domestic sewage, still more rigorous treatment technologies exist. Populations located in highly sensitive areas, such as those near fragile estuaries or once pristine lakes (e.g., Lake Tahoe, which is located between California and Nevada) require additional treatment to extract nitrogen and phosphate compounds that can contribute to eutrophication. These additional treatment systems can be very costly.

So, how can global progress in sanitation be measured? The statistics are striking. In the past century, cholera has become a rare disease and certainly is no longer an epidemic in the Americas, the United Kingdom, Europe, and parts of Asia, predominantly because of our advances in sanitation. From the standpoint of surface-water health, figure 2.1 tells a remarkable story. Dissolved oxygen concentrations are a good reflection of stream and ecosystem health. Monitoring of these levels over the past 120 years shows the demise of the Rhine River in Europe, the Thames River in England, and the New York Harbor in the United States. As sewage-removal abilities improved, human decimation of surface waters increased as societies dumped large volumes of untreated wastewater into nearby water bodies. As sewage treatment technologies were implemented in the mid- to late 1900s, however, the health of the surface water slowly improved and many rivers have largely recovered today.

The left axis of figure 2.1 shows the actual concentration of dissolved oxygen in the water. The reader will immediately note that O_2 is not very soluble in water, again, only at about 8 to 10 mg/L or parts per million at the temperature of most stream waters. Yet this low oxygen level is sufficient for many aerobic organisms to thrive. The right axis is more important from the standpoint of the environmental impact of waste, since it shows the relative level of dissolved oxygen actually present in the stream. During the decades of high pollution with organic matter, the dissolved oxygen levels dropped to 20 to 40% of natural levels and greatly impacted the survival of aerobic life forms.

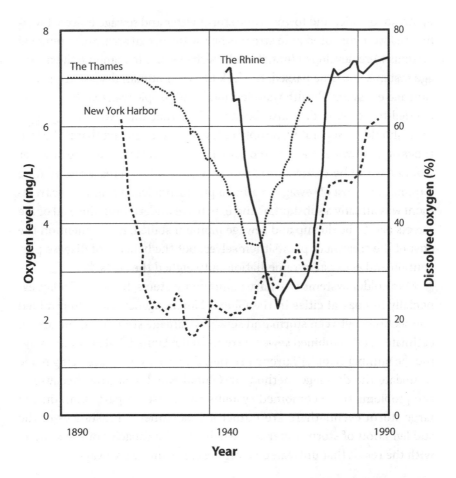

FIGURE 2.1. The effects of sewage inputs on dissolved oxygen levels in three major waterways of Global North countries. *Source*: Meadows et al. 2004. Used with permission from Chelsea Green Publishing Company.

The Next, Necessary Steps for Sewer Infrastructure

While we have had considerable success in treating our sewage waste, and in turn cleaning up our rivers, several necessary improvements remain. One is a legacy of early sewage-removal systems, the first of which was simply the sloped middle of the streets. As populated areas developed, however, combined drainage systems were installed that removed storm water and sewage from population centers. Since water flows downhill, this worked well—drinking water systems were located uphill and upstream of cities and towns while storm water and sewage flowed down-hill and were disposed of in streams below the urban area. Since removal systems were developed first, it made perfect sense to combine the drainage systems. As time passed, the link between human disease transmission and ecosystem health was identified, and people realized that sewage needed to be treated. It also became clear that the two drainage systems should be separated. Simply put, people recognized that if the two types of wastewater were mixed together, a much larger volume of water would have to be treated. This would significantly increase the volume handling and cost of sewage treatment plants. Implementing sewage treatment was already a substantial and expensive undertaking since all of the streets had to be dug up and sewage piping installed, not to mention the cost of the treatment plants themselves, but the benefits of disease prevention and ecosystem remediation outweighed the costs.

Many older systems still have combined systems, however. Today, especially in coastal cities in the Global North, agencies still find mixed connections between storm and sewage drainage systems. There are an estimated 700 combined sewer systems in the United States alone. Imagine the number in older European cities. Extensive efforts are being made to update the drainage methods in Global North countries because of the problems these combined systems can cause. In particular, during large storm events there are numerous, documented instances of the sudden flood of storm water overwhelming the capacity of the system, with the result that untreated sewage escapes into waterways.

The Next, Necessary Steps for Sewage Treatment

Although the Global North has spent billions of dollars to install full-scale wastewater systems that produce water that is cleaner than the stream water receiving it, many wastewater systems still need to improve their treatment methods, especially the removal of phosphate. Phosphate is excreted from humans and contained in synthetic detergents. It is a major nutrient that contributes to eutrophication in streams, lakes, and estuaries and it is one of the primary causes of dead zones at the outlet of major rivers into oceans. Phosphate pollution can be treated through technology that is added onto standard wastewater treatment plants.

There are two main ways to eliminate phosphate: enhanced biological phosphorus removal or chemical precipitation, in which relatively inexpensive iron salts are added to the wastewater. Removing phosphate in the treatment process not only prevents the possibility of eutrophication from the wastewater effluent, but has the added advantage that if the composted sludge from the treatment plant can be used as fertilizer, the additional phosphate makes the compost an even better fertilizer. Of course, if the wastewater itself is to be applied directly to farm lands, phosphate removal would be avoided during the treatment process.

What do we need to do next to address the problems caused by low pollutant concentrations from pharmaceuticals, antibiotics, and natural and animal growth hormones? Unfortunately, we do not currently have a satisfactory answer. The first advice, as you have probably heard from your local pharmacist recently, is not to flush any medication down the toilet but instead bring it back to the store for proper disposal or dispose of it in your garbage. This is an important step toward removing many pharmaceuticals from the aquatic environment. But what about the many chemicals that we produce and consume and then excrete and flush into our sewer system? We have treatment options for the easy pollutants, essentially the low-hanging fruit of proper waste disposal. But there are pollutants in the effluents of even the state-of-the-art wastewater treatment plants that are wreaking havoc on the environment, such as the pharmaceuticals and endocrine-disrupting compounds mentioned above. The concentrations of these chemicals are not excessive and many are as low as parts per million and parts per billion. Yet even these low levels can be harmful to the environment. All of my life I have been told to flush left over prescriptions down the toilet but most pharmaceutical compounds are not treated, adsorbed, or degraded to any degree even in the most state-of-the-art sewage treatment plants. These chemicals pass straight through the plant, untreated and not significantly degraded by microbes, and into the water of the receiving stream.

We certainly need more research on advanced wastewater treatment systems that can target these compounds. It is doubtful that we will have an answer in the near future, but the answer will likely involve an advanced biological treatment at significant extra cost.

While the amount of health-care chemicals entering our streams from our homes is considerable, this does not even begin to account for the massive inputs of hormones and antibiotics from compact livestock operations and aquaculture.

One of the first publications to capture the extent of pharmaceuticals and hormones in our wastewater and as a result, eventually in our "natural" waters, was a landmark publication by the U.S. Geological Survey (USGS) in 2002 (Koplin et al. 2002). Scientists at the USGS measured the concentrations of 95 compounds in 139 streams across the nation. Most of the selected monitoring sites were in streams suspected of being polluted with these compounds. The compounds most commonly found in the selected streams were two types of steroids (hormones), caffeine (a stimulant), a disinfectant (triclosan), a fire retardant, and a partially degraded form of a detergent. The measured concentrations, generally in parts per billion, rarely exceeded state or federal limits, but their presence raised concern about the chemicals exiting our wastewater treatment plants. Numerous investigations have confirmed these findings, although work is ongoing to determine what the impact of these chemicals may be on organisms and ecosystems.

So, who are the culprits that mainly release these compounds? Again, industry, which serves us, and each of us in our everyday lives. Almost all pharmaceutical and health-care products that you consume are eliminated from your body and flushed down the toilet. Other industrial sources, all of which potentially release chemical agents to receiving streams, include drug manufacturing, agriculture, animal medicines (especially from large livestock operations), and aquaculture. Our current wastewater treatment plants are highly variable in their ability to treat or remove these types of compounds. Once in a stream or other body of water, some of these chemicals degrade, others do not.

What do we know about the effects these chemicals are having on the environment and biota living in these waters? The simple answer is that these effects are substantial. European scientists have conducted studies on many rivers receiving treated wastewater from highly populated areas and have found gender imbalances in fish populations. Some fish populations have proportions of females as high as 70%. Studies have shown that these changes happen during sexual development in fish eggs and have also noted many anatomical changes.

What about behavioral changes in biota living in these waters? This has been the subject of many recent investigations. Various fish species, chosen because their size makes them easy to study, show changes in hyperactive, antisocial, and aggressive behaviors. Some even get the "munchies" when exposed to water containing trace levels of antianxiety drugs! This can alter species' mating behavior and entire predator–prey relationships, thereby upsetting the natural balance of populations in river water. Notably, these abnormalities have been shown to occur at very low parts per billion concentrations. This raises the question: What are the long-term effects of these concentrations? In addition, what are the outcomes of lower concentrations and of mixtures of drugs and similar agents that we are not even able to detect? Finally, how does this tainted fish tissue affect us when we eat it? Many questions remain unanswered.

In addition to disrupting ecosystems, the presence of drugs in the environment can raise an entirely different problem. You most certainly have heard of drug-resistant microbes, often referred to as "superbugs," which are tremendously dangerous to human health. One of the easiest ways to grow or develop a superbug is to expose a microbial population to low concentrations of antibiotics, such as the triclosan found by the USGS in 139 streams, over long periods of times. In just a few seasons, you too can have a superbug in your nearby stream!

Great Success, but at What Costs?

As we have seen in this chapter, in Global North countries we have solved two big wastewater problems: (1) our rivers are no longer open sewers and we now adequately treat the sewage that does spill into them, and (2) through good sanitation practices we have eliminated disease epidemics as well as most diseases associated with human waste. But at what cost? As noted in chapter 1, treatment plants are paid for by state and federal grants (i.e., your tax dollars) and by municipal bonds that you help pay off in the form of monthly water and sanitation fees. The cost of a sewage treatment plant for a city with a population of 100,000 is on the order of $40 to $50 million, spread out in monthly water bill payments costing only tens of dollars. For comparison purposes, recall from chapter 1 that the cost of a gallon of clean drinking water ranges from less than 1 penny to 10 to 20 cents depending on the technology needed at a particular location, while the cost to treat a gallon of sewage is in the range of one to tens of cents. If you are wondering about the aggregate costs at your residence, these numbers are fairly linearly scalable, and are easily recognizable on your local water bill. And remember that clean water depends on both water and wastewater treatment.

River Pollution

How clean are our rivers? As discussed earlier, pathogenic microorganisms are not a major issue in rivers because of the constant water flow and the short life spans of these organisms outside of a warm-bodied host. However, chemical pollutants can be an issue in rivers because not all chemicals that enter a stream or lake stay dissolved in the water. Many pollutants, such as dissolved metals and organic compounds, have a high affinity for particles in the water. Because of erosion and natural water runoff, most of our lakes and rivers contain plenty of particles that have been washed in from their surroundings. When dirt particles and pollutants come together, a variety of chemical and physical attraction mechanisms cause the pollutants to stick to the particles. Particles present in natural water can differ greatly in size and buoyancy and some dirt particles clump together into aggregates. Depending on the amount of mixing in a water body, different particles will settle out in different regions of the waterway. Large, relatively heavy particles will settle out first in well-mixed regions of the river or lake, while very small particles settle into the calmer, deeper regions of the water body. Considerable concentrations of pollutants that are absorbed by these particles therefore settle out as well. In principle this seems desirable. Pollutants attached to particles in the water or in the sediment are far less available for living organisms to consume or for causing toxic reactions, particularly when they settle out. Over time, as more and more dirt is washed into the water, layers of sediment accumulate in the bottom of the water system. Thus, it would seem that the pollutants have been trapped in the sediment and will not harm the biota. However, several processes, some natural, some unnatural, can result in the resuspension of polluted sediments.

There are several aquatic animals that survive by eating and sieving through the sediment. The resuspension of sediment by biota is referred to as bioturbation and it can reexpose the aquatic ecosystem to very old pollutants, particularly those that do not degrade, such as metals and chlorinated hydrocarbons (i.e., DDT [dichlorodiphenyltrichloroethane] and PCBs [polychlorinated biphenyls], carcinogenic and endocrine-disrupting chlorinated organic compounds). Similarly, storms that result in increased mixing of the system can reexpose and resuspend these polluted sediments. Both of these potentially harmful processes are unavoidable.

One unnatural, anthropogenic resuspension event is the dredging of shipping channels. Shipping channels are naturally the deepest portion of the water body and as noted earlier, they contain the smallest of the settled out sediment particles. Small particles have a very high surface-area-to-diameter ratio and carry large quantities of organic matter, two important parameters that tend to increase the amount of pollutants fixed to each particle. When these particles are resuspended during a dredging event, a high concentration, shock load of pollutant can be released to the ecosystem. The possible effects of numerous ports around the world maintaining their shipping channels can be considerable. Dredging operations are highly regulated to minimize the release of pollutants, but releases do occur. The decision to dredge a waterway is always contentious, and a detailed cost–benefit analysis is involved—the cost and risk to the ecosystem from exposure to pollutants versus the benefit of ensuring continued viability of a waterway important to commercial transport. There is seldom a clear answer to this dilemma.

One wound that we have imposed on rivers is dams. In the name of flood control and power generation, we have constructed approximately 75,000 dams on virtually every river of any size in the United States. Dams greatly interfere with the self-cleansing ability of rivers since they store large volumes of water that contain nutrients that would be washed out of the system. Generally, various forms of life will grow until they run out of the least available nutrient. For farm crops, the limiting nutrient is nitrogen, usually in the form of ammonium nitrate (NH_4NO_3), although phosphate (PO_4^{3-}) is also important and added to many crops. The limiting nutrient in freshwater systems can also be nitrogen or phosphate, depending on the system. A class of aquatic organisms known as blue-green algae (cyanobacteria), however, can actually take very un-reactive nitrogen from the atmosphere (N_2) and convert it to a form of nitrogen that is readily available to other plants (ammonia [NH_3], ni-trites [NO_2^-], and nitrates [NO_3^-]) in the water. The problem is that when nutrients such as nitrogen and phosphate are retained within the water behind hydroelectric dams, algae grow rampantly, particularly with warm summer temperatures. When the population of organisms at the base of the food chain increases too quickly, they thrive during the sun-light hours through photosynthesis and produce lots of dissolved oxygen for other aquatic organisms, but the production of oxygen crashes at night, when all nonphotosynthetic organisms respire and consume oxygen instead. As a result, there is simply not enough dissolved oxygen at night to satisfy all of the biomass in a nutrient-rich aquatic system. Thus, this matter dies and is suspended, causing murky bodies of water. This entire process is referred to as eutrophication. To put this in perspective, if you were to lower your hand into a eutrophic waterway, you could not see it deeper than 6 to 12 inches in the water! While dams are great for flood control and inexpensive power generation, they can kill a river biotope. The more dams on a given river, the worse the problem.

Lake Pollution

Polluted lakes are a far more complicated problem to solve because of the high water retention time as compared with rivers, which rapidly carry water to the ocean. Some of our most polluted water systems are lakes. The water in the lake is polluted, the sediment is polluted, and the an-nual cycling of nutrients from the sediment makes natural remediation almost impossible and certainly long term. For example, Lake Erie is the fourth largest of the five North American Great Lakes. In the 1960s, Lake Erie was declared dead because of nutrient overloading from human and animal waste and fertilizers that have high concentrations of added ni-trogen and phosphorus. These ingredients produced eutrophication based on too much biological growth, as discussed above. Because of this rapid productivity, many portions of the lake eventually begin to lack sufficient oxygen for the natural biological species to survive, a condition referred to as hypoxia. Scientists predicted that it would take many decades to address the sources of pollution affecting Lake Erie and additional decades for the lake to recover. Fortunately, local citizens, governments, and industries acted aggressively, and while the lake still has many problems, including a seasonal dead zone in its western basin, the overall health of Lake Erie has greatly improved. In contrast, Lake Onondaga in Syracuse, New York, did not receive this level of early, concerted citizen action and remains perhaps the most polluted lake in the United States. In the late 1800s and early 1900s the lake was a prized source of recreation but as the city grew, the lake became polluted with sewage, high concentrations of salt from the production of soda ash, and mercury from a chlorine production plant. Despite decades of remediation efforts, the water quality has barely improved.

Status of Water Pollution

So, what is the state of waterways in Global North countries (meaning North America, Europe, and some parts of Asia)? These countries have made reasonable accomplishments in cleaning up their waterways. The chemical industry is generally highly monitored and fined when pollution events occur, and accidental releases are less common now than in decades past. Industry has adjusted to governmental mandates, and today most industries take pride in their environmental records. New industrial facilities have pollutant abatement planned into their system from the initial plant design. Older facilities have had to retrofit their plants through what is called end-of-pipe technology, in which pollution abatement units are added to the end of processing lines. In fact, the development of pollutant technologies is now a large, job-creating, and profitable business sector. Pollutant abatement can be achieved by very simple technology such as merely adjusting the pH of wastewater to meet the NPDES standards of the Environmental Protection Agency (EPA) or precipitation of a metal out of the water column and into a holding basin prior to the release of wastewater. More complicated and expensive technologies include oxidation-reduction reactions, ion-exchange resins, and activated carbon filtration. All of the technologies used today are well developed, proven, and widely accepted by industry, the EPA, and also many environmentalists.

However, accidents will continue to happen with any human-controlled operation such as an industrial plant. One of the riskiest components of many industrial facilities is storage lagoons, where large volumes of aqueous waste or waste slurries are contained behind earthen structures. The largest volume of waste ponds contain material with relatively high concentrations of toxic metals resulting from mining operations and ash produced by coal-fired power plants. Again, the problem with metal waste is that it never degrades and is usually retained within the sediments of the receiving rivers and lakes. There are currently approximately 600 coal-fired power plants in the United States (http://www.sourcewatch.org /index.php/existing_U.S._coal_plants). These plants result in hundreds of ash disposal or storage sites, approximately half of which were landfills and the other half were waste ponds. Each year roughly 140 million tons of ash waste, containing arsenic, lead, selenium, and hexavalent chromium are produced. Too often, especially during rainy seasons in the East and Midwest, the earthen structures built to retain the metal-containing ash fluidize when the soil contains too much water to be stable, and hundreds to thousands of square yards of ash-bearing water flow downhill to towns and streams. Cleanup is difficult if not impossible. A notorious example was the December 22, 2008, disaster at the Kingston Fossil Plant in Tennessee, where a dike holding back decades' worth of coal ash failed and flooded residential areas with more than a billion gallons of toxic-metal containing waste. Yet another release of toxic arsenic-containing coal ash occurred in 2014 in the Dan River in North Carolina by Duke Power. Interestingly, and sadly, coal ash is not subject to federal regulation, and state laws governing coal combustion waste disposal are usually weak, especially in big coal-using states. So the message is, while we have made great strides in cleaning up prior water pollution events, citizens and governments must stay vigilant.

But regulating industry and treating industrial wastes, wastewater was, and will be, the easy piece of the overall puzzle. Individually, industrial plants that produce waste are referred to as a *point source,* meaning that you can point to the effluent pipe and identify the point of emission. There are many other pollution sources that fall into the category of *nonpoint* sources, where pollution stems from a large runoff area such as a farm field, large mining operation, or a leaching landfill. The atmosphere is a huge potential nonpoint source of pollution in terms of air pollutants that get mixed into precipitation; acid rain is a prime example. The biggest threats to rivers and lakes today are nonpoint sources of pollution that are largely created by human actions. We are loving our water bodies to death by having riverfront properties too close to the water. Only rarely in these types of housing situations do municipal septic systems exist; normally, individual septic tanks and field lines are the preferred and economical choice. These systems are generally effective, but when the field lines are located too close to the waterway, nutrients, especially nitrogen-containing compounds and phosphates, leach into the surface water. The larger the number of houses located too close to the water, the more leachate is released and the greater the overgrowth (eutrophication) of the body of water.

Likewise, farming operations, while certainly necessary, will produce the same problem when located too close to a water body. Some states require buffer zones around sensitive bodies of water that consist of tens to hundreds of feet between the edge of the farm and the neighboring water body. These buffer zones are proving effective in removing nitrogen and phosphate nutrients before the surface runoff reaches the stream.

But even with great efforts to remove pollutants such as nutrients from our bodies of water, one problem still occurs throughout the world, even in the cleanest of rivers. Remember that rivers are very effective at washing pollutants out of the system and into the ocean. As these relatively nutrient-rich waters, sometimes still laden with nitrogen and phosphate from farming operations and poorly performing sewage treatment plants, enter inlets to the ocean (estuaries, bays, and large river effluents), the intersection of these water systems thrive with life—and not always in a good way. Many freshwater–ocean interfaces have historically been good fishing grounds, but as the human population has grown and our inputs of nitrogen and phosphate have increased in the rivers feeding these estuaries, high seasonal algal growths have followed. As algal colonies die at the river–ocean interface, they sink to the relatively shallow bottom, which is cut off from atmospheric oxygen supplies, and decay, consuming most of the dissolved oxygen in the water. This is followed by a rapid death of all oxygen-dependent aquatic life. While a few of these dead zones occur naturally, there was a rapid increase in the number of dead zones and the severity of the low oxygen content in various bodies of water during the nineteenth century.

Currently, there are an estimated 200 dead zones around the world, most of them concentrated on the coasts of the United States and Europe. Because of the devastating effects on the fishing industry, efforts are being discussed or attempted around the world to reduce nitrogen and phosphorus inputs from farming and sewage treatment or from the lack of sewage treatment plants entirely. The largest dead zone in the world is in the Baltic Sea. Depending on the season and the year, it can cover 120,000 square kilometers (46,000 square miles), or approximately half the size of the United Kingdom. While scientific investigations have documented that this area has experienced dead zones for around 8,000 years in the deep regions of the sea, human development has extended the problem to the shallow waters. The Baltic Sea is, however, a serendipitous success story. The collapse of the Soviet Union and its state-funded fertilizer programs resulted in large reductions of nitrogen inputs. Since the late 1990s, dead zones in the shallow waters have all but disappeared. In the United States, improved sewage treatment in recent decades has similarly shrunk the dead zone near Long Island Sound off New York City to a considerable degree. However, efforts to reduce other dead zones have not shown great progress to date, most likely because of the storage of nutrients in sediments that are recycled each year.

Also off the coast of the United States are noted dead zones in the Chesapeake Bay and the Gulf of Mexico. The Chesapeake Bay is well known for its abundant and popular seafood, especially the blue crab. Generations of fishermen have experienced declines in the numbers of catches as a result of shifting dead zones in the bay. Many efforts have been brought to fruition, and in some cases implementation of nutrient-reduction programs are in practice, but the dead zone persists and will for years to decades into the future because of continued inputs of nitrogen and the storage of nutrients in the sediment.

Another dead zone that is reported in the news every year is that at the mouth of the Mississippi. The Mississippi River is the largest river in North America, starting in northern Minnesota and slowly working its way southward for 4,070 km (2,530 mi) to the Gulf of Mexico. Thirty-one U.S. states and two Canadian provinces are contained in its watershed, which covers more than 3,220,000 km^2 (1,245,000 mi^2). Along the way from its headwaters to its outlet, the Mississippi collects tons of suspended matter (dirt), giving the river its muddy color, which contains suspended nutrients from farming operations and from sewage treatment plants. As these nutrient-rich waters reach the Gulf of Mexico, the waiting sea life—cyanobacteria, green algae, dinoflagellates, coccolithophores, and diatom algae—explodes. These organisms' rate of productivity is so high during the warm season that a tremendous amount of algae rot late in the summer and their decomposition consumes all or most of the dissolved oxygen that is required for the survival of higher forms of life such as fish and crabs.

Historically, the dead zone in the Gulf of Mexico was contained to an area of approximately 5,000 km^2 (1,930 mi^2). The size of the dead zone has been increasing since the 1980s, however, and has now reached a surface area of 15,000 km^2 (5,800 mi^2), which is larger than the state of Connecticut. It has been estimated that as much as 74% of the nitrate in the Mississippi is flushed in from farms and another 10% results from wastewater, mostly from sewage treatment plants. In the most severe case, the dead zone extended from the mouth of the Mississippi to the Texas–Mexico border. Not surprisingly, this massive dead zone has had a negative impact on the Gulf of Mexico's fishing industries. Task forces have come and gone, and more studies will be conducted, but the obvious way to reduce the growth of the dead zone, and shrink it to its historical and eventually natural proportions, is to reduce the amount of nutrients released into the watershed. But this will not be an easy, or even achievable, task given that the Mississippi River drains the American Midwest breadbasket.

Accidents Will Continue to Happen

While we have made great strides in cleaning our rivers and lakes, and in controlling routine industrial releases, accidents will continue to occur. We are able to prevent a large number of them with tighter monitoring, but accidents are only part of human nature. The goal is that with more rigorous laws and industry inspection these unintended releases will be minimized. On January 9, 2014, a possibly avoidable chemical spill occurred in West Virginia when Freedom Industries released 7,500 gallons (28,000 liters) of 4-methylcyclohexanemethanol (MCHM) into the Elk River. Ironically, MCHM is a chemical used to remove various pollutants from coal prior to burning it in coal-fired power plants. The incident affected the water supply for 300,000 residents for at least a week. Symptoms of MCHM exposure include headaches, diarrhea, dizziness, reddened and burning skin, nausea, vomiting, rashes, and itching skin. By mid-January numerous lawsuits had been filed against Freedom Industries and in classic corporate fashion, Freedom Industries filed for Chapter 11 bankruptcy on January 17.

In another incident, a pipe carrying wastewater owned by Duke Power in North Carolina released 39,000 tons of coal ash solution containing arsenic at 14 times the level deemed safe by the EPA into the Dan River. Potential exposures from the spill included water contact and the ingestion of fish from the affected reaches of the river. Arsenic exposure can result in skin, bladder, and lung cancer. And again it should be noted that there are an estimated 600 EPA-identified coal-ash waste sites in the United States alone, storing millions of tons of coal ash contaminated with toxic metals including arsenic, lead, and selenium. This is another hidden cost of Big Coal. What are we to do with all of this toxic waste?

These reports are just another day in the life of industry. Many of these human-caused events can be prevented, but accidents will continue to happen nonetheless, and we must be prepared for them.

Extending Our Existing Successes to Developing Countries

Even with the above challenges still facing wastewater treatment in Global North countries, we have made great advances in treating our waste, cleaning our rivers, and reducing disease from improper sanitation. But what about sanitation in Global South countries? There are between 2.5 billion and 3.5 billion people without access to basic sanitation, or 4 to 6 of every 10 people on Earth. The dumping of untreated sewage might be the crime of the last century in some countries, but it remains the crime of this century in other nations. Chapter 1 also noted that the cost of giving everyone in the world access to clean drinking water and basic sanitation is estimated to be about $35 billion to $70 billion per year from now until the year 2025. Although we are making substantial strides in bringing safe drinking water to all of the world's residents, the challenge of basic sanitation is far greater.

While all humanitarian efforts are noteworthy, Global North countries, through agencies like the United Nations and the World Health Organization, have certainly made our share of mistakes in aiding some Global South countries. One historic and fundamental problem to past approaches is assuming that solutions that have worked in the Global North are appropriate in other countries. An example of such multinational support, backed by the World Bank, was the paved BU 364 highway in 1984 to Brazil's newest state Rondonia that actually increased habitat destruction and development. Another classic example of a U.S. funding disaster, which resulted in a 400% overrun of costs, was the Inga-Shaba Extra High Voltage DC power transmission from the Inga Dam complex at the mouth of the Congo River to the mineral fields in Shaba that fell under the corruption of the Mobutu government, and later rebels. There are the many errant, internationally funded irrigation schemes, most notably one in Sudan that was initiated in the 1920s and subsequently faced drought. I could go on and on, but you get the point. Foreign aid is good, even great, but must be implemented wisely, must be made relevant to facts on the ground, and most of all must be sustainable, especially with regard to construction and day-to-day maintenance of water and wastewater treatment facilities.

Cost-Effective Options for Global South Countries

So, while we have excellent, advanced technology for treating sewage wastewater in the Global North such that exiting water is often even cleaner than the receiving stream water, these approaches require massive construction efforts, daily energy operating costs, and highly trained personnel to keep the treatment properly functioning, all of which may make these facilities difficult to implement in Global South countries that face greater capital constraints. Wastewater treatment facilities in the Global South must therefore be sustainable with respect to low cost of the plant, simplicity of design and maintenance, low or preferably no energy (electricity) use, and low or preferably no chemical use. The wastewater treatment needs of small populations can be met with a variety of latrine options. Larger populations can effectively be serviced with two sustainable and relatively low-cost alternatives, waste stabilization ponds (WSPs) and constructed wetlands (CWs).

WSPs are large, shallow water-retention structures that are easy to construct and that rely completely on natural processes performed by bacteria and algae to oxidize and stabilize biological waste. They can be composed of only one pond or several connected in a series. Domestic wastewater enters at one end of the pond and very slowly moves to the other end before potentially moving into another pond. The more ponds in series, the more controlled the system and the more the waste is treated. Since these ponds are simple and do not require any energy inputs to operate, the treatment or oxidation of the biological waste is very slow. For example, water retention time in a modern plant in the United States is on the order of 8 hours, while retention times in WSPs in other areas can be as long as 30 days for complete treatment of biological waste and removal of most nutrients and pathogens. WSPs are commonly used in many rural areas of Global North countries that lack adequate electricity for mixing motors.

Many geographic locations currently in need of wastewater treatment also experience water shortages. Fortunately, water from treatment systems such as WSPs typically still contain the necessary nutrients for crop irrigation and fertilization, and when they are operated properly, dangerous pathogens will have been removed. Reuse of treated wastewater is also commonly applied in aquaculture in the Global South. WSPs can therefore serve beneficial, dual purposes: treatment of biological or sewage waste and as a source of nutrient-rich irrigation water.

For low-lying coastal or delta areas, CWs are common. They even exist at interstate rest facilities in remote parts of the United States. These facilities use river reeds to extract nutrients out of wastewater and rely on wetland-dwelling microbes to oxidize the organic waste.

Two questions are worth asking at the close of this chapter: Could our rivers have been spared from the abusive pollution resulting from modern civilization? And were there any concerns raised about these practices even back in ancient times? Sewage and pollution problems did not arise overnight, instead they built up over decades, and in some cases over centuries. One of the best analyses of the industrial revolution of the mid-1800s is by Rosen (2003). There have been mixed and conflicting interpretations of historical events of the period but it is possible to identify two influential forces: the practice of the common law inherited from England that was thought to favor industry and the use of antinuisance law that tended to side with citizens fighting against all types of pollution. After analyzing 46 pollution-related nuisance cases from the mid-1880s in Massachusetts, New York, New Jersey, Pennsylvania, California, Illinois, Ohio, and Texas, Rosen suggests that courts at the time tended to favor nuisance law settlements over more traditional forms of pollution regulation. The same courts, however, failed to recognize more modern forms of industrial pollution as a nuisance. Basically, technology and development has historically outpaced the courts and our legislative system.

The North American Free Trade Agreement (NAFTA) between Canada, the United States, and Mexico in the early 1990s provides some examples when considering how environmental protection is implemented by law today. Concerns were expressed at the time about the export of U.S. jobs, worker conditions in Mexico, and the "racing" of industry to the country with the most lax environmental laws. Some of these concerns came true, especially regarding the metals and petroleum industries. But no one predicted the forthcoming workers' rights issues and environmental disasters that rapidly emerged in Asia's recent industrial revolution. Asia, specifically India and China, have strict environmental laws concerning waste emission. The problem is that local officials do not adequately enforce the laws for a variety of reasons, particularly corruption. China has produced in only one decade the same level of pollution that the United States experienced in a century. Based on what little information has leaked from China, it is clear that the air, rivers, and estuaries are extremely polluted. Could the United States have helped prevent the disasters we experienced from occurring in Asia? We will have to leave this question to future historians. The good news is that citizens of Asia are adamantly speaking out and demanding action.

This chapter and the successes noted in chapter 1 present humans' greatest environmental success story—the ability to supply clean drinking water and adequately treat and sanitize our wastewaters. Although it has taken centuries to achieve these goals, it offers us a direction to attack today's environmental problems. In contrast to upcoming chapters, there was no corporate villain or greed-based decisions to overcome, each of us, in masses, was the source of the problem. We are an inseparable part of Earth's environment and must accept our responsibility as its stewards.

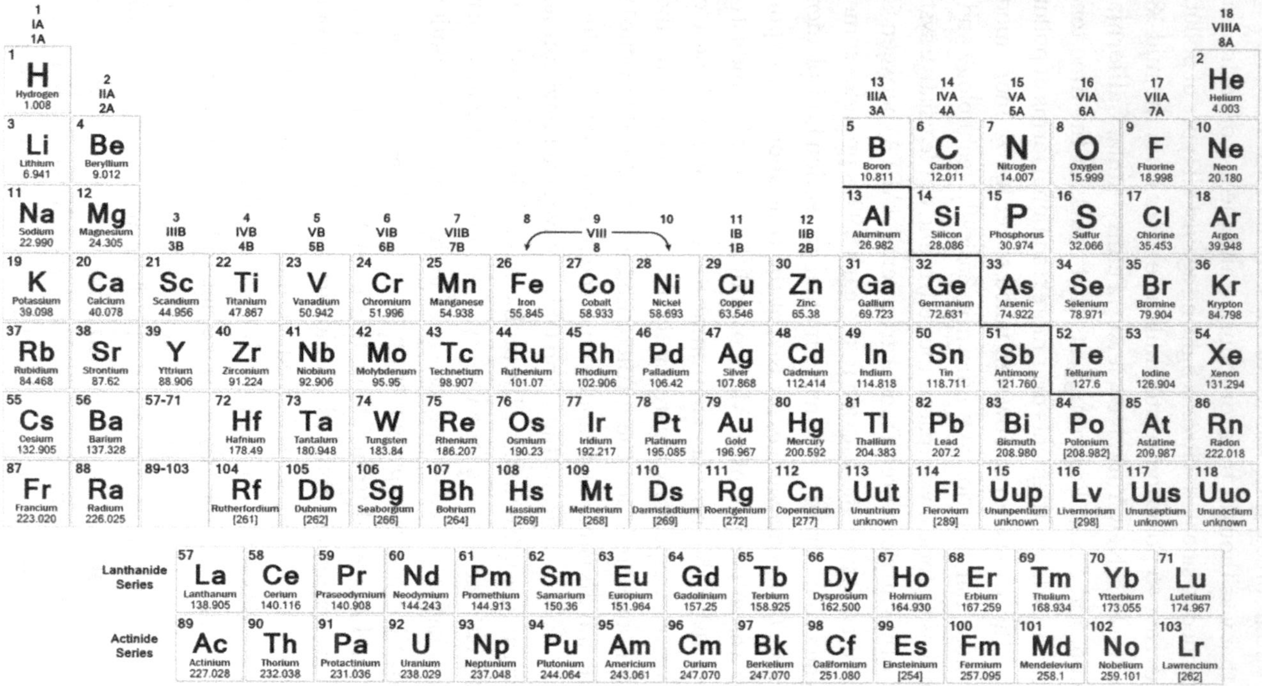

FIGURE 3.1. Periodic Table of Elements

THREE

The Removal of Anthropogenic Lead, and Soon Mercury, from Our Environment

*About seven years later I was given a book about the Periodic
Table of the Elements. For the first time I saw the elegance of
scientific theory and its predictive power.*

—SIDNEY ALTMAN

Much of chapters 1 and 2 focused on dealing with relatively straightforward biodegradable, organic human waste. For the remainder of this book, and especially in the next two chapters, we will discuss the environmental effects of specific chemical pollutants. This chapter focuses on lead, which has the chemical symbol Pb from the Latin word *plumbum,* and on mercury, which has the chemical symbol Hg from the Greek word *hydrargyrum* (meaning "water and silver" or "liquid silver").

Lead and mercury are both classified as metals as are all of the chemical elements to the left of the solid jagged line in the right-hand portion of the Periodic Table (figure 3.1). Many books have been written dedicated to the Periodic Table. As noted in the quote by Nobel Prize winner Sidney Altman, the Periodic Table is the defining organizational system for chemical elements. It is viewed by many scientists as the greatest organizational chart of all time. There are many ways of explaining the structure of the table, such as by the number of protons (subatomic particles with a positive charge and negatively charged valence electrons each element has), by each element's reactivity, and by the interpretations of the Schroedinger wave function equation. All of these characteristics complement each other and are in fact interrelated, so they therefore arrive at the same organization of the elements.

Our organizational focus in this chapter will be a bit different from that of a chemist and is more in line with that of a biochemist. First, I will provide an overview of the history of the discovery of lead and mercury, then their use in modern society, a bit of biochemistry about their toxicity, and then how we have used them irresponsibly throughout human history. But today we are on the cusp of mostly eliminating human-generated sources of these, and other toxic heavy metals, from our lives. As I will discuss, exposure to lead in water and that produced by industry is well under control in most Global North countries, and we are finally starting to control mercury emissions from industry. This is indeed another welcome success story. Now is time to help the Global South control their toxic metal exposures.

The Elements: A Basic Overview

First, it is important to note the "life" elements gathered in the top right portion of the table, most notably carbon (C), nitrogen (N), oxygen (O), sulfur (S), and phosphorus (P), and the misplaced hydrogen (H) located in the upper left. These elements are shown in a bold font and are major components of all known life forms. Many other elements are instead neutral to life, especially in the small concentrations that exist on Earth. Another grouping of elements could well be considered "death" elements because they can be toxic at certain doses, but even some of these are in fact necessary for life in trace quantities; e.g., chromium (Cr), arsenic (As), and selenium (Se). Elements such as cadmium (Cd), lead (Pb), and mercury (Hg) are harmful at any concentration.

As shown in the Periodic Table, elements are said to exist in their elemental state; in other words, they have all of their electrons, which is equal to their number of protons. Elements therefore do not bear an overall charge. But in nature, most elements, including lead, exist as minerals (with the notable exceptions of pure geologic veins of copper, silver, gold, and a few others). Elements held within minerals exist as more complex compounds that contain: (1) a positively charged ion of the metal element since it has given up electrons, and (2) some negatively charged ion such as chloride (Cl^-), oxygen (O^{-2}), or phosphate (PO_4^{3-}) (as well as others) that have taken up electrons. This assemblage usually either forms a relatively stable mineral or is easily dissolved in water.

In water, all metals exist as cations, positively charged ions that have a variety of water molecules associated with them (known as waters of hydration); these can result in beautifully colored solutions. Most metal cations are environmentally friendly and even essential for life, such as sodium (Na^+), calcium (Ca^{2+}), magnesium (Mg^{2+}), potassium (K^+), and others. Other metals such as cadmium (Cd^{2+}), tin (Sn^{2+}), lead (Pb^{2+}), and mercury (Hg^{2+}), however, can be toxic even at extremely low concentrations in water. But how can one metal be beneficial to life and another, even nearby on the Periodic Table, cause death? The answer depends on the chemical form of the metal to which you are exposed, how it enters the body (by inhalation or orally), and if it can cross the air–blood barrier in our lungs, the gut–blood barrier, and the blood–brain barrier.

Lead

Historical Uses of Lead

Of course, most of the 91 metal elements exist in nature, in varying quantities, while the remaining few have been created only in nuclear reactors. Lead is present naturally on Earth in approximately 94 minerals, mainly in ores mixed with zinc (Zn), silver (Ag), and copper (Cu). Galena (lead sulfide, PbS) is the most common lead-containing mineral. Lead is rarely found in its pure metallic form. Metallic lead is instead recovered from ores through smelting, with several million tons of lead produced annually. Common uses of lead today are lead-acid batteries, bullets and shot, weights, and as key components of solders, pewters, metal alloys, and radiation shields. Previous, unfortunate uses of lead included pure lead piping (most notably by the Romans). Tetraethyl lead was previously used in gasoline as an anti-knocking component. Various salts of lead were once used to produce white, yellow, and red paints. Japanese geishas even used lead carbonate for painting their faces white. Lead was also deployed as an insecticide. The point of mentioning all of these examples is to emphasize that while toxic forms of lead have been with us for centuries, we have only recently come to understand this element's toxicity.

Today we know that almost all forms of lead are toxic. The metal form, as shown in the Periodic Table, is toxic through inhalation or ingestion and adversely affects all organs in the body. Aqueous salts of lead [chemical forms that are soluble in water such as lead nitrate $Pb(NO_3)_2$ and others] can be readily absorbed into the blood system after ingestion. Lead mainly targets the nervous system as both a short- and long-term neurotoxin that inhibits the transmission of chemicals between neurons or nerve connections. The passage of lead across the blood–brain barrier results in the instant death of affected neurons in the brain. Ingested lead salts, also known as ionic forms of lead, can pass the gut–blood barrier and wreak havoc on internal organs. The mode of toxicity of the lead ion, and most toxic metals, is by reversibly binding to essential proteins in an organ, especially in the liver and kidneys. This binding changes the shape of the protein and interferes with its ability to bind in normal biochemical reactions. In biochemistry, shape, polarity, and binding ability of molecules, including important proteins, are everything, as further illustrated in chapter 4 in its discussion on endocrine disruptors.

Although lead naturally occurs in many minerals, humanity's first widespread exposure to the metal form of lead was during ancient Greek and Roman times. While the first smelting of lead occurred as far back as 8000 B.C.E., the metal had little application in weaponry because of its softness. However, lead smelting reached its maximum during the height of the Roman Empire. The Greeks and Romans used lead extensively in water containment systems and piping because of its malleability. Though not known at the time, use of these lead pipes resulted in Pb^{2+} ions being released into the drinking water. One enjoyable but regrettable application of lead was the heating of wine in lead pots, which sweetened the wine through the formation of lead acetate (basically, lead vinegar). This certainly increased Romans' blood and brain levels of the neurotoxin and some even say it contributed to the fall of the Empire because of the resulting insanity of its leaders. A few Roman writers of the time, such as Pliny the Elder and Vitruvius, actually noted the negative health effects of lead in water piping (Book VIII of *De Architectura*). Fortunately, southern Europeans today have improved their winemaking processes and no longer need the added taste from lead acetate. The use of lead in water piping continued for centuries, however, and was even used in early, and some later, American waterworks.

As previously mentioned, the 1900s saw a variety of less-than-ingenious uses of lead in piping solder, leaded paint, and leaded gasoline. All of these sources have been eliminated in the United States today, but it took until as late as the mid-1980s. The need for lead solder for sealing piping joints was easily eliminated with the common installation of PVC and other plastic piping in homes. Leaded paint, especially lead white, presented a hazard only when it flaked or peeled off the painted surface. Anyone with small children knows that they love to put everything in their mouth, and much of this flaking material and the lead it contained was easily absorbed into their bodies. Lead paint is particularly hazardous when removed through sanding since the resulting lead dust is almost unavoidably inhaled. But, our historical exposure to lead from solder and paint sources was minimal in comparison to the egregious act of adding lead to gasoline.

Corporate Greed and Creation of an Anti-Knocking Fuel

Early automobiles were fueled by a variety of combustible liquids depending on what was available at the time and at a particular geographic location. The early fuels tended to pre-combust in the gasoline piston chamber. This tendency to pre-combust is measured in the octane rating that relates a particular fuel combustion ability to octane, an eight-carbon modern gasoline component. As combustion engines became more powerful, the need for more properly combustible fuels became apparent. The key to developing more powerful engines was higher combustion pressures and higher-octane fuels. This combination resulted in problematic engine knocking, the same sound that your engine makes when it continues to run after you turn off the ignition, but in the pre-combustion of low-octane fuels that happens during normal engine operation. Thus, an anti-knocking chemical was needed that would also improve fuel efficiency. Yes, fuel efficiency was a concern even back in the 1920s! Thus, corporate America stepped up to fulfill this need, but only in line with its own best interests and profits.

In a special report in *The Nation* published in a March 2000 issue, Jamie Kittman chronicles one of the most striking stories of corporate lies, government manipulation, and greed ever recorded. And, if General Motors', Du Pont's, and Standard Oil of New Jersey's (now ExxonMobil) efforts were not bad enough on their own, Kittman claims that their profit schemes "have provided a model for the asbestos, tobacco, pesticide and nuclear power industries, and other twentieth-century corporate bad actors, for evading clear evidence that their products are harmful by hiding behind the mantle of scientific uncertainty" (page 1 of the Internet version).

I encourage you to put down this book for a moment, Google "The Secret History of Lead: Special Report by Jamie Kittman," and read the entire article. It is relatively long but well worth the read. I'll only briefly summarize it here. Research in pursuit of a suitable anti-knocking compound began prior to the 1920s. As it turned out, race car drivers at the time already used alcohol in their engines with great success in avoiding engine knocking and improving fuel efficiency. Even in 1917, Kettering and Midgley, two bad actors discussed further below, recognized ethyl alcohol (also known as farm alcohol, since it is derived from plants) as the best anti-knocking fuel. But Corporate America had a problem with ethyl alcohol. If any farmer could make it in the bathtub, how could industry make a profit by producing it? Enter tetraethyl lead (TEL), commonly referred to by industry as "ethyl" in order to draw attention away from the fact that it contained lead and instead use a play on words to conflate the two-carbon compound, ethyl, with the alcohol—ethanol.

TEL was first discovered by the German chemist Karl Loewig in 1854. At the time, it was considered of little use given its known deadliness; in fact, it is even more hazardous than the lead acetate in Roman wine. Even in 1854, TEL was known to cause hallucinations, madness, spasms, palsies, difficulty breathing, asphyxiation, and death. But several decades later, industry executives realized that more importantly, TEL could be patented as an anti-knocking additive to gasoline and yield large profits to the patent holder of an anti-knocking gasoline additive. The early 1920s saw a flurry of activity concerning TEL: data manipulation, death of plant workers, and concerns vocalized by academic researchers and the U.S. Public Health Service about human exposure to TEL once the new proposed gasoline formula was implemented. But in the end,

Corporate America won, and for 60 years we used a metal known by the Romans to be toxic. Further, the organic form of leaded gasoline was extremely toxic and the contamination it caused was widespread—cars sprayed lead all along the nation's roads and farm equipment emitted it over acres of crops. After the introduction of leaded gasoline, government data showed an increase in lead-blood levels, especially in children, and it has been estimated that 5,000 Americans died annually from lead-related heart disease prior to the phase out of leaded gasoline. As Kittman summarizes in his article, "Profit at any cost."

Phasing Out Lead from Gasoline

So, how did we finally remove lead from gasoline? As it turns out, by indirect action. Smog became such a problem in major U.S. cities that by the early 1970s the U.S. Environmental Protection Agency (EPA) had implemented measures to reduce emissions from automobiles that served as precursors to smog as well as emitting other toxins. The major chemicals in smog formation were, and are, hydrocarbons and nitric oxide (NO), which react with the air to produce nitrogen dioxide (NO_2) and free radicals. Since incomplete combustion also creates carbon monoxide (CO), a toxic gas, the EPA also sought to prevent its emission from automobiles. To control these emissions, catalytic converters were first installed on automobiles in 1975 and have been improved over time. Today, catalytic converters convert NO and NO_2 back to regular, atmospheric nitrogen (N_2), convert CO to CO_2 (a nontoxic but global warming–causing gas), and oxidize the remaining hydrocarbons to CO_2. The installation of catalytic converters in automobiles, now required in most Global North countries, has greatly reduced the presence of smog-producing gases in congested cities in these countries.

So, what does this have to do with lead in gasoline? An unexpected side effect of adding catalytic converters to cars to reduce air pollution was that lead from tetraethyl lead plated out on the metal catalysts in catalytic converters, rendering them useless. The removal of lead from gasoline was not the initial intention, but was instead an added benefit of smog reduction. It is also interesting to note that this was the first paradigm change in the cost of environmental remediation. Instead of taxing corporations for pollution or cleanup, consumers paid for the change in the cost of the catalytic converter in their new automobiles.

As a side note, it should be mentioned that tetraethyl lead manufacturers did not immediately stop producing TEL. Instead, they shifted sales efforts to Global South countries, much like how the tobacco industry has shifted its pursuit of profits to these same countries. Successful removal of TEL from all gasoline in all countries is slowly being achieved, but much more work is still needed in Asia and Africa.

So, without TEL in our gasoline, what kept engines from knocking after the late 1970s? Fool me once, shame on me. Fool me twice, shame on my government regulators. There were still profits to be made from anti-knocking additives, and a few bad actors in industry immediately pushed a new synthetic chemical, methyl tertiary butyl ether (MTBE). MTBE contains extra oxygen that can be used to increase the octane content of gasoline. Problem solved: no knocking engine, more fuel efficiency, and continued profit for industry. Like TEL, however, MBTE was also environmentally harmful: all gas station tanks eventually leak, and unlike gasoline, MTBE is soluble in water, so it readily moves into groundwater. MTBE gives a foul taste to drinking water even at concentrations of 5 parts per billion. Limited data on MTBE risk are available for the low concentrations found in groundwater, but it is unlikely a carcinogen. Today, many groundwater sources are contaminated with MTBE and therefore unusable for procuring drinking water.

Finally, in the early 2000s, U.S. states started requiring the replacement of MTBE with ethanol, the same chemical that Kettering and Midgley acknowledged back in 1917 as being among the best anti-knocking fuels. In fact, in the early twentieth century, ethanol was considered the fuel of the future. After decades of industry gaining profit while emitting lead into our environment and poisoning untold numbers of people, the safe and simple compound ethanol with its anti-knocking properties is finally in our tanks. Simply put, it took nearly a century to implement a solution we knew about all along. But this is not to say that ethanol production from massive amounts of corn grown in our Midwest bread basket, at the expense of growing food crops, is the optimal answer.

The Benefits of Eliminating Lead

In contrast to leaded gasoline, the phaseout of lead paint was a less controversial and delayed process. This was largely because alternative paint formulas were developed in the mid-1800s that did not result in profit losses. By the 1940s, titanium dioxide pigments were standard in the United States, and in 1971 President Nixon signed into law the Lead-Based Paint Poisoning Prevention Act (LBPPPA), which defined the permissible lead content of paint to be 1%, later lowered to 0.06% through amendments. In 1992, Congress passed the Residential Lead-Based Paint Hazard Reduction Act, which established paint with a 0.5% lead content as hazardous, necessitating control measures. Lead paint does remain in some very old houses even today, however.

Since the removal of lead from paint and gasoline the level of lead found in children's blood has dropped dramatically. Lead still persists on most U.S. roadsides, but as you move even slightly away from roads, lead levels drop rapidly to natural background levels. Remember that lead is a natural element and that trace concentrations are common in soil. Lead concentrations in U.S.-grown foods are considerably lower than in the agricultural products of countries still using or that recently used leaded gasoline. For all practical purposes, Global North countries have solved the lead crisis. Now, our efforts must shift to the rest of the world and finally eliminate all profits from leaded gasoline throughout the world.

But I should end this section on a cautionary note. Many systems in the United States still contain lead piping or lead solder in the iron pipe joints. One of the most common lead-containing items in water distribution systems is knows as a "goose neck," a lead joint that connects the main distribution line to the small distribution systems. This is normally accommodated for by adding trace amounts of phosphate to essentially eliminate lead dissolving (or pipe corrosion) from these sources into drinking water; this has been common practice since the 1800s. But Flint, Michigan, city managers and elected officials chose to ignore science, chemistry, and advice from their consulting engineering firm when they switched water sources from Lake Detroit to the Flint River, which had a different water chemistry. By attempting to save $50,000 on the new water system, they poisoned Flint residents with lead that leached from

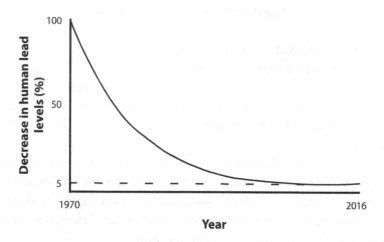

FIGURE 3.2. Generalized decrease in human lead levels over time

the pipe distribution system and permanently damaged the entire piping system that will now cost tens of millions of dollars to replace. Hopefully someone will go to jail for choosing to save a few dollars over solid scientific evidence known since the 1800s. It is unknown how many more systems like the Flint water distribution system are in the United States.

The major remaining source of lead in our drinking water is the lead goose neck contained in many, or most, of our distribution system. But as noted earlier in this chapter we have reduced the human lead burden by 95% since the 1970s.

But as the data show our success has stalled in the last two decades, and we can't seem to overcome the remaining 5% pollution level (see figure 3.2). I attribute this small remaining lead exposure to our water distribution systems and leaded solder in households. While it would be easy to say, just replace the culprits, goose necks, we have little to no idea where they are in our distribution systems. Can you imagine the excessive cost of digging up every street in every town containing a central distribution system? This concept brings up a central theme in pollution management and abatement, cost–benefit analysis, illustrated in figure 3.3.

Let's take human lead exposure as our example. Removing lead from gasoline and paint was relatively easy as compared to digging up every street in the United States. The removal of lead from gasoline and paint

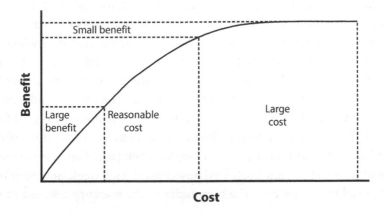

FIGURE 3.3. Conceptual benefit–cost relationship

yielded a huge benefit, a 95% reduction in human lead exposure, for a reasonable cost. But removal of the remaining 5% of lead exposure, from lead piping in our water distribution systems, is simply not cost effective to do at one time. Certainly, over time, as routine or slightly increased maintenance schedules allow, we will eliminate the remaining 5%. For now, a simple and relatively low-cost in-line filter on your kitchen faucet will remove your remaining exposure to lead.

Mercury

Another metal that humans have overused throughout history is mercury. Mercury's chemical symbol is Hg and it is located very close to lead (Pb) on the Periodic Table. Mother Earth gives us mercury in 17 mineral forms. The most prevalent is mercuric sulfide (HgS), more commonly known as cinnabar. An extremely rare metal in the Earth's crust, mercury can nonetheless be concentrated in mineral deposits at up to 2.5% by weight. Mercury can occur in a variety of chemical states, most commonly in the Hg^{2+} cationic state. With heating of certain metals or industrial processing it exists as elemental or liquid mercury, while microbial processing of ionic mercury (Hg^{2+}) produces a deadly neurotoxin: methylmercury.

Although mercury is rare in the Earth's crust, we still need to be concerned with its presence because very small amounts of mercury are harmful and even deadly. One natural source of mercury over which we have no control is volcanoes, which have contributed about half of the existing atmospheric mercury through historical eruptions across the globe. The other half of atmospheric mercury has been produced by human actions. The major anthropogenic source of mercury is coal-fired power plants, which are responsible for about 30 to 40% of the atmospheric burden. This could certainly increase, however, since China is currently constructing about one coal-fired power plant each week. Several minor industrial processes emit the remainder of the mercury released to the atmosphere.

The average citizen comes into contact with mercury by using thermometers, barometers, manometers, blood pressure gauges, dental filling amalgams, and fluorescent lamps. Society is rapidly converting to dyed alcohol thermometers and thermistor-based devices. Although it will still take several years, fluorescent light bulbs will eventually be completely replaced with more energy-efficient, light-emitting diodes (LEDs). Liquid mercury used in gold mining to extract the gold from fine particles in dredging operation was previously a major source of environmental releases in Global North countries, and it still is in many other countries. Important progress has been made in chloralkali plants, historically one of the major mercury-using industrial processes, which today are being converted to mercury-free operations in the Global North. Chloralkali plants produce the many millions of tons of chlorine gas and sodium hydroxide required by industry. Prior to the mid-1980s, chloralkali plants used liquid mercury as an electrode, with the unintended result that each plant released a few hundred pounds of mercury each year. In the mid-1980s, new plants started using a diaphragm method that was mercury-free. Considerable efforts have been made to reduce mercury emissions from the nine major chloralkali plants remaining in the United States. The PPG (now Axiall Corp.) plant in Lake Charles, Louisiana, even converted to the more modern and environmentally friendly membrane cell technology in 2009. My sources in this industry tell me that there are only two major U.S. chloralkali plants remaining that use mercury. In addition, by 2020 all European chloralkali plants are scheduled for closure. This is welcome news!

Mercury Emissions from Coal-Fired Power Plants

Now for the largest source of anthropogenic mercury: the burning of coal. The approximately 572 existing coal-fired plants in the United States together emit 48 tons of mercury each year. Asia, taken as a whole, has an even higher number of plants that in aggregate emit 1,460 tons of mercury per year. Coal use is declining in the United States, but its use is drastically increasing in many Asian countries. In 1990, coal-fired power plants in Asia contributed 28% of the total, annual mercury emitted into the atmosphere from all sources, natural and anthropogenic.

A good question to ask at this point is how did mercury get into coal? And if mercury is in coal, what other toxic compounds might be present? In fact, coal contains numerous problematic elements, namely arsenic, cadmium, chromium, lead, and selenium. In addition, thallium and uranium released from the burning of coal are major sources of human-released radioactivity. As a result, it has been noted that coal-fired power plants emit more radiation (from thallium and uranium) than do nuclear power plants by accidental releases.

So, how do these toxic elements get into coal? Coal is essentially mineralized or fossilized carbon derived from million-year-old vegetation from low-lying areas. As these areas were buried deeper and deeper over time, heat and pressure from the overall lying strata compressed the organic carbon, squeezed out water, and formed coal. Elements that existed only in trace concentrations in the original plants are now present in relatively high concentrations in the condensed material that is coal. Mercury in U.S. coal is around 0.17 parts per million. While that may still sound minor, consider the following 2011 coal production figures produced by various countries *each* year: China, 3,520 millions of tons; the United States, 993 million tons; India, 589 million tons; the EU, 576 million tons; and Australia, 416 million tons. All of this coal will be burned somewhere and will release corresponding tons of mercury into the atmosphere. The resultant atmospheric mercury will travel with wind currents around the world for several months before falling to Earth attached to particulate matter or in rain.

Environmental and Human Health Effects of Mercury

Once atmospheric mercury enters aquatic systems, a variety of reactions can occur. Ionic mercury occurs as divalent mercury or mercury with a positive ionic charge, Hg^{2+}. As with lead and all toxic metals, if ingested or inhaled, mercury has adverse effects on organs such as the heart, liver, and kidneys. This harm results from the deformation and deactivation of necessary proteins as mercury binds to sulfur atoms within proteins containing cysteine and methionine, important amino acids. Mercury is even more problematic because it can also be present as elemental mercury (Hg^0) and as organic forms like methylmercury (CH_3-Hg^+) and dimethylmercury (CH_3-Hg-CH_3). Elemental mercury, when inhaled, can be easily transported across the blood–brain barrier and is a deadly neurotoxin; yet ingested elemental mercury seems to be relatively harmless. Historical records show that elemental (liquid) mercury has been associated with human health as far back as Egyptian times and that it has been used both in medicines and as a poison. More recently, when mercury was used to cure and protect the rims of hats, its vapor toxicity was reflected in the expression "mad as a hatter."

While only inorganic forms of mercury are emitted naturally and by industry, an even more toxic form of mercury results when mercury finds its way into aquatic systems that have low oxygen content. A unique, but completely natural form of bacteria can, perhaps as a detoxification mechanism, add organic carbon to inorganic mercury to form CH_3-Hg^+ and CH_3-Hg-CH_3. But an even bigger problem is that the organic form of mercury stays neither in the bacteria nor in the aquatic system. Released into the water at very low concentrations, even parts per billion, methylmercury, like toxic metal cations, is attracted to the sulfur groups of amino acids that make up necessary proteins in aquatic microorganisms. Bigger organisms eat large volumes of these methylmercury–containing microorganisms, which in turn ties up more proteins in the larger organisms. This continues up the food chain: rotifers eat these contaminated organisms, small fish eat the rotifers, and larger fish eat the smaller fish. As the process continues; higher masses of organisms are eaten and more methylmercury is concentrated at each step up the food chain. While only parts-per-billion concentrations or less of methylmercury may be present in the water originally, these concentrations can grow to as high as parts per million in the top feeding species. This process is referred to as bioconcentration and can result in neurological damage to the top predators.

So, what happens when humans eat fish at the top of the food chain? We get considerable doses of a highly poisonous neurotoxin. There are two variations of this for fish: freshwater and oceanic species. By 1997, about 33 states had issued fish consumption advisories because of mercury contamination, mostly by methylmercury. In 2004, the EPA reported that more than half of the fish residing in 260 freshwater lakes across the United States contained mercury concentrations above the recommended level for pregnant women and small children. Infants are particularly susceptible to mercury poisoning because methylmercury has more dramatic effects on their brain development and because they lack the necessary biochemistry to excrete mercury from their bodies.

More advisories are being issued every day. In addition to animals that live high on the food chain, animals that live long lives also accumulate mercury. Whales are somewhat lower on the food chain since they mostly eat krill, but because they live for many years, they bioaccumulate some of the highest levels of mercury in the oceans. Ironically, if word spreads that whale meat has mercury concentrations exceeding recommended consumption levels, this may be what actually puts an end to the whaling industry once and for all. And readers have certainly heard the warning concerning certain types of tuna with high mercury concentrations. As Asia builds more and more coal-fired power plants, this problem will only be exacerbated. For example, the U.S. Geological Survey (USGS) predicts that the mercury levels in the Pacific Ocean will double between 1995 and 2050.

Since humans both live high on the food chain and have long lives, we are doubly affected by bioaccumulation. One should also remember, however, that fish, especially components of fish oil, have essential nutrients. So, we need to consume fish. But, given concerns about mercury concentrations, what fish are actually safe to eat? The EPA (http://water.epa.gov/scitech/swguidance/fishshellfish/fishadvisories/index.cfm) recommends the following guidelines for women and young children:

1. Do not eat shark, swordfish, king mackerel, or tilefish because they contain high levels of mercury.
2. Eat up to 12 ounces (2 average meals) a week of a variety of fish and shellfish that are lower in mercury.

   • Five of the most commonly eaten fish that are low in mercury are shrimp, canned light tuna, salmon, pollock, and catfish.
   • Another commonly eaten fish, albacore (white) tuna has more mercury than canned light tuna. So, when choosing your two meals of fish and shellfish, you may eat up to 6 ounces (one average meal) of albacore tuna per week.

3. Check local advisories about the safety of fish caught by family and friends in your local lakes, rivers, and coastal areas. If no advice is available, eat up to 6 ounces (one average meal) per week of fish you catch from local waters, but don't consume any other fish during that week.

Linking Accumulated Mercury Emissions to Human Actions

So, how do we know if much of the mercury in today's fish and in the environment in general is related to human activity? Remember that half of the mercury in the environment has accumulated for millions of years from volcanic eruptions. Geologic deposits and glaciers can be used to study the past since sediment and ice have been deposited in certain areas for hundreds, thousands, and millions of years. Digging down in a sediment deposit or in ice in a glacier is like moving back in time. The most recent deposits are at the top, with very old deposits far deep down. The actual sample is called a "core" since a coring device is used to collect it. Data from cores, taken from a Wyoming glacier by the USGS, are shown in figure 3.4.

The y-axis shows the approximate age of the cores, which dates back to considerably before the industrial revolution. Total mercury concentrations in nanograms per liter, or parts per billion, are shown on the x-axis. Overall, figure 3.4 shows a fourfold to fivefold increase in the deposition of atmospheric mercury into ice since the Industrial Revolution, clearly demonstrating human causation. Large spikes in mercury

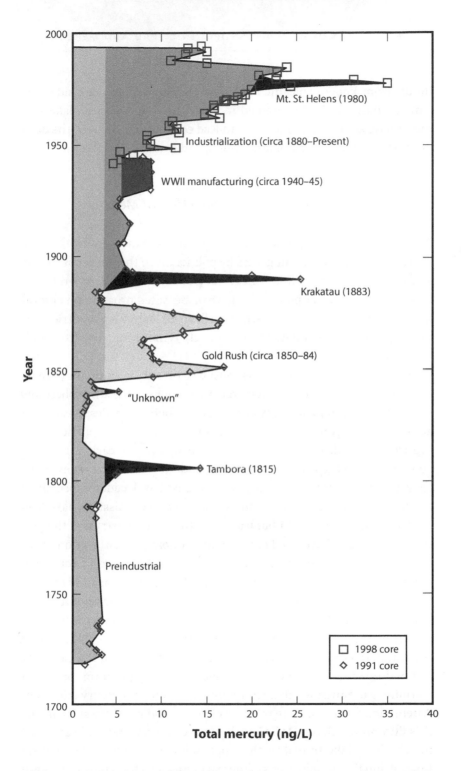

FIGURE 3.4. Profile of historic concentrations of mercury in the Upper Fremont Glacier (Wyoming, United States). *Source*: Schuster et al. 2002. Reprinted with permission from the American Chemical Society.

inputs from known volcanic eruptions are also visible. As a side note, similar studies of lead contained in European and American lake sediments have shown large increases in lead concentrations since the dawn of smelting, during the Roman Empire, and from leaded gasoline.

Phasing Out the Remaining Uses of Mercury

After decades of delay, numerous profit-driven maneuvers by the fossil fuel industry, and constant games by politicians in the pocket of the fossil fuel industry, we are finally beginning to regulate mercury emissions from coal-fired power plants. We have made noted progress on chloral-kali plants but old power plants, especially in Global South countries, desperately need to be updated to completely remove mercury from their processes.

Again, coal-fired power plants must be a priority because they are the major source of human-caused mercury emissions to the atmosphere and, subsequently, to our aquatic systems and our food chain. Responsible citizens have been concerned about coal for decades if not centuries, starting with smog deaths in London and Europe in the early nineteenth century. If only those first, concerned Londoners knew the many toxins they were taking into their lungs. For the past few decades, especially in the United States, but around the world, mercury emissions have been repeatedly in the spotlight but each time the coal industry and the politicians they fund have cited costs to the economy as justification for inaction. The 1990 amendments to the Clean Air Act listed mercury, and compounds containing mercury, as a hazardous air pollutant and required the EPA to set emissions standards for its sources. Much political maneuvering by the fossil fuel industry ensued and continues today.

During the George W. Bush administration, the EPA sought to remove regulation of coal-fired power plants from section 112 of the Clean Air Act and instead create a cap and trade regulatory program for global-warming gas emissions that would indirectly affect mercury emissions. After much federal to state transfer of power and many lawsuits against this EPA proposal, the D.C. Circuit vacated the cap and trade approach in February 2008. In reality, this approach had little to no hope of significant and timely reductions of mercury emissions from coal-fired power plants. After this court decision, the EPA was forced to redesign its approach to regulating air pollution emissions from coal-fired power plants, including mercury emissions. In the meantime, many states took the initiative in developing new standards. In 2011, the EPA proposed a rule that would require new coal-fired power plants to install pollution control technologies to reduce mercury emissions, as well as emissions of arsenic, chromium, nickel, and acid gases. Once the Mercury and Air Toxics Standard (MATS) is fully implemented, it will require existing power plants to install the necessary pollution-abatement equipment. We are well on our way to removing mercury from our industrial processes in the United States. But I will note that the technology planning to be used today was available and proved during the George W. Bush administration as early as 2005. In 2015, there was one minor setback when the U.S. Supreme Court ruled against the EPA's power plant mercury limits, since industry claimed that the EPA did not include economic costs and job loss in their assessments. But this setback was only minor, since the rule was not struck down by the Supreme Court but only sent back to the previous court for review, and as the EPA notes, most plants have already implemented the new emission requirements and pollution abatement equipment and EPA now includes economic costs in the assessment of new ruling.

But what about the rest of the world? With China completing the construction of a new coal-fired power plant each week and with Asia's existing release of 1,460 tons of mercury each year, Asian citizens now need to appeal to their governments, and they are becoming more outspoken in their activism every day as smog and toxins affect every aspect of their lives. I hope that change in these countries does not take as long as it has in the United States. In this regard, 140 nations attending the United Nations Environmental Program in Nairobi, Kenya, in 2009 agreed to begin negotiating a global treaty to control mercury emissions. Finally, in 2012, these countries, including the United States and China, came to consensus on a set of legally binding rules to reduce mercury pollution. Referred to as the Minamata Convention (after the 1956 Japanese Minamata Bay tragedy involving mercury poisoning), the treaty will require coal-fired power plants, industrial boilers, and other large industrial facilities to limit mercury emissions using the best available technology. Success at last!

As a critical next step, the world must reevaluate the appropriateness of coal as a useful fuel. When you consider the rampant harms from mining, including human death and black lung disease, the significant environmental damage from mountaintop removal and acid-related drainage, the energy costs of transporting coal to power plants, the toxic metals emitted into the air from burning coal and contained in the resulting toxic fly ash waste (most notably mercury, arsenic, chromium, and cobalt), the release of toxic fly ash during storage dam failures, the cancer-causing very fine particulate emissions that enter our lungs, the sulfur oxide emissions that cause acid rain, the fact that coal-fired power plants release more radioactivity than the commercial nuclear industry, that coal powers huge contributions of carbon dioxide gas (is a major source of global warming), and the associated, extensive, future economic costs including carbon dioxide sequestration, is coal really worth it? Is coal really less expensive than renewable alternatives when you add in human suffering, death, and extensive environmental degradation? We will return to the subject of coal in chapter 9, in which we discuss climate change.

The Legacy of Lead and Mercury

The negative impacts of our previous actions will be with us for as long as humans exist, because metals do not degrade. The toxic metals that we have released will cycle in our food chain and continue to haunt us until Mother Earth slowly buries them in the deep ocean or lake sediments, a process that will take centuries.

One of my first jobs, at 13 years of age in 1970, was pumping leaded gasoline at a full-service station. I was always perplexed as to why there was a sign on the pumps boldly stating that there was lead in the gas. Why was lead added to gas? Wasn't lead toxic? Where did it go when it was burned in the engine? If a 13-year-old of average intelligence can ask these questions, should not the chemists, engineers, and managers of the time at Standard Oil (now ExxonMobil) have been asking the same questions? Similar logic can be applied to historic industrial uses of mercury and the politicians supported by Big Coal. Both metals have been known for centuries to be neurotoxins. Why did industry continue to use them? Why did our government allow this? My generation carries a lot of harmful legacy chemicals in our bodies, and I certainly blame my slow mental moments on my early exposure to lead and mercury. We clearly know who to blame for our toxic metal burden. When will the corporate leaders and corporations ever be held accountable?

The good news is that after decades of efforts we are well on our way to avoiding these toxins in order to protect today's children. Though it may take a few decades of persistence, the successful phaseout of lead and the ongoing removal of mercury from our industrial processes demonstrates that when we coalesce around a common goal, we can overcome even significant political hurdles and government–industry collusion.

FOUR

Elimination of Chlorinated Hydrocarbons from Our Environment

If we are going to live so intimately with these chemicals—
eating and drinking them, taking them into the very marrow
of our bones—we better know something about their nature
and their power.

—RACHEL CARSON, *SILENT SPRING*

In many ways, it's hard for young people today to go back in time and understand the power of Rachel Carson's statement. For readers who lived through the 1950s and 1960s, you witnessed rivers on fire, streams that served as open sewers, the previously discussed death of Lake Erie, industrial smokestacks belching black smoke full of who knew what, and many other environmental atrocities. It was easy to point to these highly visible environmental offenses, but could these chemicals be permeating our very bone marrow, or worse? Industry and government were less concerned, but increasing occurrences of cancer began to create fear among the population.

Today we know more about cancer, and we have shown that many industrial chemicals do cause cancer. We also now have a whole new worry: the industrial endocrine-disrupting chemicals present in many of our consumer products that operate under the same biochemical mechanism that *Silent Spring* warned about for birds exposed to DDT. Surprisingly, a few bad actors in industry issue the same defensive reaction today . . . no harm from our chemicals! And they may be correct for some of the new chemicals, but it is more likely that many will be harmful. For the most part, we have eliminated many of the historically notorious "poster chemicals" from use, even global use. This is a great success story that can be a lesson for our next steps—dealing with endocrine disruptors.

This chapter will focus on both our past success and our next steps sur-rounding the topic of endocrine disruption—a broad class of natural and synthetic chemicals that can or are interfering with our natural hormone-producing systems. We will start with a little organic chemistry and then examine the historical and first-noticed problem of pesticides interfering with the reproduction of birds of prey, such as the American bald eagle. Next I will address endocrine disruptions in humans by chemical contam-inants in our daily foods and consumer products. And finally I will report on our successes and future efforts for predicting endocrine-disrupting behavior of new chemicals prior to their use and release into the envi-ronment.

A Brief Explanation of Organic Chemistry

First, in order to understand this chapter, I need to teach some of the basic principles of organic chemistry, the bane of college sophomores in majors such as chemistry, biochemistry, and biology and most students in health professional schools. Students normally endure a full academic year of organic chemistry, but I will expose you to only a few pages, the real meat of the subject. So, bear with me, as this information is essential to understanding important facets of the dangers facing our environment.

To begin with, what makes organic chemistry different from other chemistry courses? How are organic compounds different from other molecules? Simply put, organic chemistry is the study of any chemical composed primarily of carbon (C) and hydrogen (H), in some cases with the addition of nitrogen (N), oxygen (O), sulfur (S), and a few other ele-ments in small ratios. The simplest of organic compounds is methane (CH_4), the main product of hydraulic fracturing (fracking) for natural gas, a hot topic in the news these days. The chemical structure of methane is shown in figure 4.1. In each of the chemical diagrams included in this chapter, the gray-black sphere always represents a carbon atom and the light white-gray spheres represent hydrogen atoms.

Starting with methane, carbon atoms and other elements can be added to make organic compounds as complex as our DNA. There are literally millions of known organic compounds, ranging from chemicals that have been studied for decades to a few completely unique compounds with

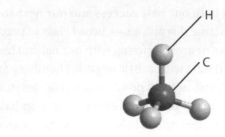

FIGURE 4.1. Methane

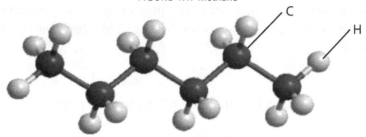

FIGURE 4.2. Hexane

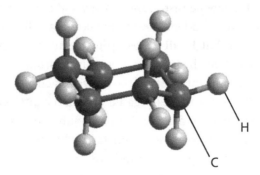

FIGURE 4.3. Cyclohexane

no known use. For illustration purposes, we'll use a six-carbon compound to discuss the many classes of compounds.

The simplest six-carbon compound is hexane, shown in figure 4.2.

This is called a straight-chain compound since it has a zigzag chain of connected carbon atoms. Additional carbon atoms can be added as vertical side chains to any of the carbon atoms. Two carbon atoms can share just one bond or up to three bonds between them, known as single, double, and triple bonds.

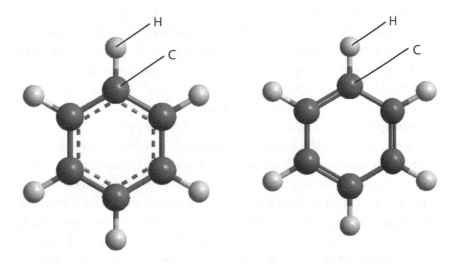

FIGURE 4.4. Benzene

Organic compounds can also exist as rings, such as cyclohexane, the straight-chain carbon depicted in figure 4.2 with the end carbons connected to make a ring, as shown in figure 4.3.

Many sugar molecules that are a staple of our diet have some version of this basic structure. Alternatively, and of great importance in organic chemistry, is the aromatic ring. Unlike cyclohexane, an aromatic ring has double bonds between alternating carbons. Aromatic rings with six carbons are known as benzene molecules, the basis of aromatic chemistry. There are several ways of representing benzene; two of these are shown in figure 4.4.

The basic structures shown in figure 4.4 can be modified by chemists, and in many cases by microorganisms in nature, to add a variety of functional groups or additional elements to the chemical structure. These groups include alcohols (an OH attached to one or more carbons, including my favorite, ethanol), ethers, esters, ketones, aldehydes, amines, and amides. You need not worry about exactly what these are; what is important to know is that they give different reactivity to organic molecules— i.e., how the specific chemical will react in a variety of reactions with other chemicals. For example, plain benzene is a known carcinogen, but an OH attached to benzene makes phenol, an antiseptic, and a COOH attached to benzene makes benzoic acid, a benign compound used to preserve food,

Now for the main point: history has shown that when you add one or more chlorine (Cl) atoms to an organic compound, especially an aromatic ring, the resulting chemical is usually very harmful to the environment and can be toxic to life. Several of these are worth mentioning, most notably the class of compounds known as the "dirty dozen" persistent organic pollutants (POPs) which include aldrin, chlordane, dieldrin, dichlorodiphenyltrichloroethane (DDT), endrin, heptachlor, hexachlorobenzene, mirex, polychlorinated biphenyls (PCBs), polychlorinated dibenzo-p-dioxins (dioxins), toxaphene, and dioxins and furans. These chemicals, and others, are referred to as persistent because they degrade very slowly, if at all, under normal environmental conditions. Those that do degrade or partially degrade, produce similarly toxic compounds that can also cause environmental problems. DDT is probably the poster pollutant of the dirty dozen. I will mostly focus my narrative on DDT, although many of the dirty dozen chemicals act similarly in the environment and in biological organisms.

DDT

As can be seen in figure 4.5, dichlorodiphenyltrichloroethane (DDT), has the requisite chemical structure of the problematic dirty dozen, including two benzene rings and five chlorines. Although first synthesized in 1873, DDT did not find its way onto the mass market as an insecticide until 1939. At the time, DDT was used with remarkable success in controlling malaria, typhus, and other insect-vector diseases, with no significant consequences for human health. In fact, DDT saved millions of lives during and after World War II, and Paul Müller, a Swiss chemist, was awarded the 1948 Nobel Prize in Physiology for his work on disease applications of DDT. After the war, DDT quickly found its way into agriculture and households for insect control and was also applied more globally to control insect-transmitted malaria. On average, more than 40,000 tons were used each year, with a total global production of 1.8 million tons produced since the 1940s. Today, only a few thousand tons are produced for controlled purposes.

DDT kills insects extremely effectively by chemically forcing open the sodium ion channels of insect neurons, causing the insect to die through

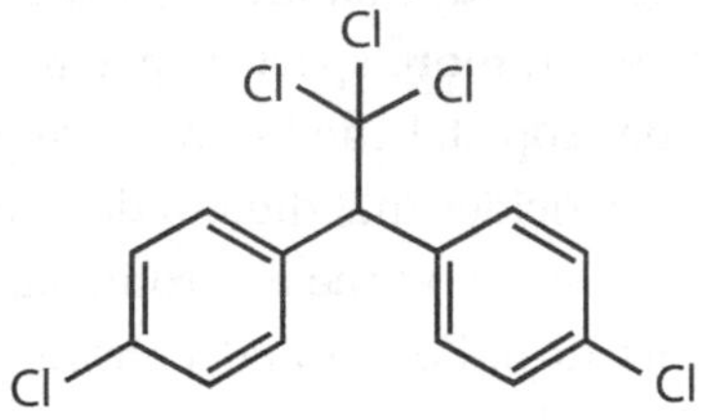

FIGURE 4.5. Chemical structure of DDT.
Hydrogens are not shown.

the instantaneous firing of all neurons. During World War II, the lack of noticeable human health problems from DDT was repeatedly asserted, even when soldiers and citizens were doused with DDT until they were white from the powder. Today, however, DDT's contribution to long-term health problems has been demonstrated, including human diabetes, development disorders, and cancer.

Catalyzing Concern About DDT and Banning Its Use

Starting in the 1940s, scientists raised concerns about the long-term safety of DDT. As usual, industry, academic researchers, and the government disagreed on facts concerning the hazards posed by DDT to human health and the environment. The nail in the coffin for DDT came in 1962 with the publication of Rachel Carson's *Silent Spring*, although it took the Environmental Protection Agency (EPA) another 10 years to finally remove DDT from the list of acceptable insecticides, which occurred in 1972. Restrictions on use followed in the United States and abroad as other nations slowly became aware of the hazards posed by DDT and acted to ban its use. Finally, after decades of debate, the United Nations Stockholm Convention of 2004 outlawed several persistent organic pollutants, including DDT. The convention has been ratified by 170 countries, with the most notable exceptions being the United States, Russia, Italy, Saudi Arabia, and Ireland. Today, India is the only country that still manufactures large quantities of DDT that it uses extensively for agricultural applications.

In the early days of insecticide development and use, a broad-spectrum approach was relied on, meaning an insecticide killed almost all insects.

As time went on, we figured out that killing pollinating species was not a good idea, and as a result, more species-specific insecticides, like the ones used today, were developed. In addition, it was previously considered a good idea to make insecticides that did not degrade. These insecticides would instead stay in the soil, and the environment, for a long time; this allowed concentrations to build up and continue to kill problematic species. We now know to be suspicious of chemicals that accumulate in the environment, and particularly of chemicals that build up concentrations in the human body, as I will address later in this chapter. Most of today's insecticides are therefore made to degrade after a given time. But unlike the history of tetraethyl lead discussed in chapter 3, there was apparently no suspicious intent by the makers of chlorinated pesticides such as DDT; only after the fact did some manufactures try to defend their products as researchers questioned its long-term implications. These chemicals originally looked like a complete win–win situation for everyone, unless you were an insect, of course.

The biggest controversy surrounding the banning of DDT was its historic use in fighting malaria. This topic is beyond the scope of our discussion, and the length of this chapter or book, but it is well presented in a recent book by David Kinkela, *DDT & The American Century* (2011). Dr. Kinkela describes therein the vicious arguments between environmental groups and international health organizations. Some health-care advocates claimed, and continue to claim, that the banning of DDT doomed millions of children per year in Global South countries to death from malaria. Others claimed that insects have developed a tolerance to DDT and that there were other, safer, insecticides that could be used to control malaria. Still others noted the destruction of ecosystems by DDT. In the end, DDT was banned in almost all countries. There is no easy conclusion to be drawn about the history of DDT, and the debate will certainly continue for decades. Today limited use has been internationally approved in malaria-infested areas. This seems to be an effective, targeted approach.

DDT as an Endocrine Disruptor in the Environment

The story of how chemicals, especially DDT, can act as endocrine (hormone) disruptors is a long sordid one, and several important and nonobvious concepts have to be understood along the way. To begin, let us revisit a concept introduced in chapter 3 on metal toxicity: bioconcentration. Many chemicals readily bioconcentrate under normal environmental conditions; this is particularly true for certain organic pollutants. Pollutants with very low solubility in water, typically less than parts-per-million concentrations, are said to be hydrophobic which literally translates to "water-fearing." These chemicals are also lipophilic, translated as "fat-loving," so they naturally concentrate in our fat deposits. DDT and its dirty dozen relatives all fall into this category, and can quickly bioconcentrate in ecosystems. An excellent example of this is from a very early study on DDT bioconcentration in wildlife in the Long Island Sound off New York City, published by Woodwell et al. in 1967 (see table 4.1).

Data in table 4.1 are arranged, from top to bottom, by increasing trophic or food-predator level in the ecosystem food chain. Plankton are

TABLE 4.1 DDT Bioconcentration Data from Long Island Sound

Medium	Concentration (parts per million)*	Concentration factor	Investment return on $1 in 10 years at different trophic levels (stocks)
Water	0.00005	—	$1
Plankton	0.04	800	$800
Silverside minnows	0.23	4,600	$4,600
Sheephead minnows	0.94	19,200	$19,200
Pickerel (predatory)	1.33	26,600	$26,600
Needlefish (predatory)	2.07	41,400	$41,400
Heron (feeds on small aquatic animals)	3.57	71,400	$71,400
Herring gull (scavenger)	6.00	120,000	$120,000
Osprey egg	13.8	276,000	$276,000
Merganser (fish-eating duck)	22.8	456,000	$456,000
Cormorant (feeds on large fish)	26.4	528,000	$528,000

*Data are from Woodwell, Wurster, and Isaacson (1967).

eaten by minnows, who are eaten by needlefish, who are eaten by herons, and so on. The second column of data shows the DDT concentration in parts per million in each species. Shown in the third column is the concentration factor, which demonstrates the amount that the concentration of DDT in a body has increased from the DDT concentration in the water. For example, a concentration factor of 10 means that the chemical has increased in concentration tenfold. You will note a dramatic increase as you compare DDT concentrations in water to concentrations in cormorants, a 528,000-fold increase.

A small number of myopically profit-driven, bad actors from the chemical industry and its supporters have argued: Why worry about a few parts per million? This is a recurring theme. Making an analogy to the parts per million that result from bioaccumulation, they say, if I have a million dollars and I give away one dollar, the cost to me is insignificant; I would not miss the one dollar. In this limited financial sense this might be true, and similar arguments can be made for one part per billion and one part per trillion. But this is a self-serving and greedy argument as well as a flawed analogy when it comes to the effects of bioconcentration. To correctly apply the analogy to money, since this is something these few bad actors can certainly relate to, let us instead imagine that the dollar is invested in a healthy Wall Street stock for 10 years, a reasonable time for bioconcentration to occur in an aquatic system. Increasing return on the stock can be analogized to the increasing accumulation of DDT in organisms as you move up the food chain, which means that, for 10 years we climb 10 trophic levels. Moving up multiple trophic levels from water to a cormorant, which has a concentration factor of 528,000, would equate to receiving $528,000 from a $1 investment made 10 years previously. I think you, and critics, can clearly see the importance of bioconcentration, and the inappropriate analogy used by those ignorant of the parts-per-million concept with respect to pollution.

Still, a few bad actors continue to argue that that the concentrations are still seemingly small and should be no cause for concern. Indeed, even at the top of the food chain, we are only talking about 26.4 parts per million or $26.40 out of a million dollars. The counterargument, however, is that these pollutant concentrations are similar to, or even greater than, the chemicals they are replacing and interfering with in our bodies, meaning that chemicals that exist naturally in our bodies on which we depend may be completely replaced or destroyed by these part-per-million concentrations of the dirty dozen. We will return to this issue in more detail later in this chapter.

So, why are DDT and its chemical relatives a danger for the environment? As noted earlier, these chemicals do not degrade to any extent and instead accumulate in every medium of the environment: water, soil, atmosphere, biota, and even in Arctic and Antarctic snow. As described earlier, these chemicals concentrate up the food chain, reaching relatively high concentrations for nonnatural compounds. It turns out that their chemical shape mimics the shape and function of important biological chemicals, a fact that was either not admitted or not known by their inventors.

Such is the case for DDT, which mimics certain estrogens in birds. It turns out that as DDT production increased (peaking in 1962), so did its concentration in ecosystems and therefore in higher trophic animal species. DDT concentrations eventually became so high that they interfered with hormone regulation in birds. How? We still do not completely understand the biochemistry pathway, but the cause-and-effect relationship is very clear. In eagles, DDT interfered with the natural cycle, or production, of estrogen and other chemicals responsible for depositing calcium in the eggshell. As a result, eggshells thinned, and when the mother eagle sat on the eggs to warm them, the eggs prematurely cracked, killing the eaglet. The populations of many upper-level bird species, also called birds of prey, plummeted. Several bird species in other countries suffered the same fate, notably the white-tailed sea eagles and peregrine falcons of Sweden. As previously mentioned, this issue was brought to prominence by Rachel Carson's critically acclaimed book, *Silent Spring*, which catalyzed action in the United States. Since the banning of DDT and similar compounds, populations of birds of prey have slowly recovered and in the United States have again reached safe and sustainable numbers. Many environmentalists therefore deem the problem of chlorinated hydrocarbons to have been essentially solved, although they acknowledge that the results of the slow degradation or burial of these chemicals in our river, lake, and oceanic sediments is still pending.

Pollutants and Human Health

Now let me address human health problems caused by environmental exposure to similar pollutants. What about all of the other chemicals that we use in today's society? Could other chemicals cause problems, in humans or other species, similar to those that chlorinated hydrocarbons caused in eagles? Answering this question is very complicated and of grave concern. The first indicator scientists look for to determine whether a chemical could be problematic is its persistence in the environment or its resistance to degradation. Of course, if a chemical does not degrade, it will eventually bioconcentrate within a food web, especially if the chemical is hydrophobic. Secondly, scientists examine whether the chemical of concern looks chemically similar to an essential biological compound in our body. This second parameter is one with which we have had very little success.

Understanding the Endocrine System

In order to understand what chemicals we ought to be concerned about, we need to review a little biology and biochemistry. While we have made great strides across scientific disciplines in recent years, the biochemistry of certain systems remains an area of intense inquiry. This is especially true of the human endocrine system, which regulates the hormones that control many biochemical functions in our bodies. On a macroscale, this system is pretty easily understood. The hypothalamus, a small area in our brain, produces and releases chemicals (hormones) that regulate other chemical releases by the anterior pituitary gland. The pituitary chemicals in turn control chemical releases in the testes for males and ovaries for females. Each of these organs (the hypothalamus, the anterior pituitary, and the testes or ovaries) has positive and negative feedback mechanisms with each other that result in increasing and decreasing the output of hormones, respectively, as needed. The difficulty with understanding these processes is that there are hundreds of hormones involved. You have certainly heard of testosterone in men, and you may be familiar with estradiol and prolactin in women, since these are the only hormones present in relatively high concentrations. There are in fact many other hormones present in minor concentrations, parts per billion and trillion, and certainly additional hormones have yet to be discovered. Each of these hormones can act independently or in unison with other hormones or other chemicals to promote or block biochemical reactions in hundreds of chemical receptor sites throughout our bodies. Given that we have yet to fully understand how the endocrine system functions, how can we possibly predict what chemical will interfere with this system? To add even more complexity, not all humans are created equal in the eyes of biochemistry. Some individuals respond positively to a chemical, others negatively, and some have no response at all. This system is a scientist's nightmare! As soon as you think you have a hormone's function and reactivity figured out, you come across a set of individuals with genes that cause them to respond differently.

In the above discussion we saw how chemicals (hormones) in very low concentrations (parts per billion or trillion) dominate other chemical reactions in our bodies. Now, let us look at more common compounds, proteins, that are present in very high concentrations in our diet (in meat and beans) and in our bodies (muscles). It was once estimated that over 100,000 DNA sequences existed, based on the biochemical functions they enable. One of the first surprising findings of the Human Genome Project was that there are only about 25,000. DNA sequences produce polypeptides that are made up of individual amino acids (proteins are multiple polypeptides put together). Proteins are necessary for biochemical functions in the human body; in fact, proteins control almost everything. The fact that fewer polypeptides were found to exist in our body than was anticipated means that a single protein must have more than one function in the human body and that groups or pairings of proteins can act in combination to serve other functions. Needless to say, this greatly complicates human biochemistry, since it would be far easier to study if one protein was responsible for one metabolic step.

The real complication arises when a nonnatural chemical enters the human body and interferes with a human protein, since this one protein can have multiple functions and therefore its replacement may disrupt numerous biological processes. The distribution of charge across the protein controls protein folding and the protein's shape, which are key to protein function. The ways in which pollutants can disrupt proper protein functioning are numerous. For example, pollutants can interfere with the biochemical manufacture of the necessary protein. They can also mimic an individual protein or group of proteins and partially or completely interfere with the binding site for the protein's target molecule (the receptor). Given the individual biochemical differences that can result in individual evolutionary advantages or disadvantages, these chemical responses will not be the same in all individuals. Life is truly complicated!

To further appreciate our individual biochemical uniqueness, let us consider real-life examples of medicinal side effects. Try to name a drug that does not have side effects or that does not have different side effects for different people. Have you ever watched a TV commercial for a medicine that did not rapidly enumerate a long list of possible side effects at the end? Side effects occur from simple drugs, such as pain relievers, and from complex medications, such as medications for the acquired immunodeficiency syndrome (AIDS). And as you may know, some drugs even have dramatically different side effects from one individual to the next. Codeine, a common pain reliever, has the frequent side effect of drowsiness, but for a rare few, the direct opposite effect occurs, euphoria. Likewise, most antihistamines, which are used to control allergies and which typically produce feelings of lethargy, result in hyperactivity in some users. Another common example of diverse reactions are side effects produced by amoxicillin, an antibiotic to which an unfortunate few have an allergic reaction. In fact, allergies themselves are a sign of our genetic diversity, since many individuals are allergic to environmental stimuli and medications that are neutral or beneficial for others.

We may never be able to fully quantify the different reactions to the same chemical in different individuals caused by the diversity in the human gene pool. But what about dose—the actual amount or concentration of a chemical that enters our body? This can be key to understanding whether an adverse human health result will occur.

Dose Toxicity in the Human Body

This brings us back to the parts-per-million, billion, and trillion argument. Some extremist environmentalists would lead you to believe that a single atom or molecule of a chemical is a danger to the environment. On the other hand, certain chemical company executives would try to convince you that their chemical is perfectly safe at any concentration. This issue will be explored further in chapter 5 on risk assessment, but the key point for this discussion is well summarized in a quote from Paracelsus (a Swiss German Renaissance physician, among other professions, living from 1491 to 1541): "All things are poisons, for there is nothing without poisonous qualities. It is only the dose which makes the thing a poison" (Brainy Quote n.d.). Again, there are some classic examples of this fact from biochemistry. Selenium, used heavily in the semiconductor industry, is toxic in doses greater than 400 micrograms. Anything above this dose can lead to a disease called selenosis, with symptoms of gastrointestinal disorders, hair loss, fatigue, irritability, and neurological damage. Trace quantities of selenium, however, are necessary for several cellular functions. Most notably, selenium is a component of antioxidant enzymes and in enzymes that regulate thyroid hormones. So selenium exposure, depending on its concentration, can be either harmful or in fact necessary. A more complex example is vitamin D, a group of eight fat-soluble compounds that act as antioxidants and are necessary in limited amounts for many other biochemical functions. But overdosing on vitamin D has also been known to occur, since our bodies do not have rapid or effective ways of excreting fat-soluble compounds.

But what about dose toxicity with respect to endocrine disruptors, the subject of this chapter? Dose determines everything, but it is also very poorly understood. For example, say that we know with certainty that a chemical is involved directly or indirectly with the functioning of our endocrine system, such as an estrogen supplement. Does one molecule of the supplement act exactly like one molecule of human estrogen? Does it take more or less of the supplement (dose) to get the desired effect? Herein lies the problem. In the case of most endocrine disruptors, we are still fairly clueless about the appropriate ratio or concentration needed to elicit the desired response for a specific function. But a one-to-one concentration is a good place to start for comparison purposes. If any pollutant in my body is at the same concentration as a necessary hormone, I would certainly be concerned.

Some values gleaned from the literature are shown in table 4.2.

Natural serum concentrations in healthy, human blood of a few well-known hormones are shown in the left-hand column. The importance is that these are parts-per-billion concentrations, evidencing the fact that

TABLE 4.2 Hormones and Pollutants in Human Blood

Natural endocrine serum concentrations	Bioaccumulated serum pollutant concentrations
Estradiol, 20–750 parts per billion (ppb)	Bisphenol A (BPA), 0.6–2.5 ppb
Progesterone, 4–250 ppb	Mendiola et al. (2010)
Prolactin, 0–20 ppb	Takeuchi and Tsutsumi (2002)
Testosterone, 28–110 ppb (F), 1.5–7 ppb (M)	You et al. (2011)
Pagana and Pagana (1998)	PCBs, 2.9 ppb
	Dirtu et al. (2010)
	Phthalate esters, 0.29–5.9 ppb
	Hogberg et al. (2008)
	Phytoestrogens in soy beans, 1,040 ppm
	Phytoestrogens in soymilk, 3,000 µg/100 g

parts-per-billion chemical concentrations are very important in our bodies. How do these concentrations compare to pollutant concentrations also typically found in normal, hopefully healthy, humans? A few data points are shown in the right-hand side of the table. BPA (bisphenol A), a widely suspected endocrine disruptor, along with PCBs (polychlorinated biphenyls) have been measured in human blood at parts-per-billion and parts-per-trillion concentrations, respectively. Phthalate esters, a pollutant common in many plastic products, is even higher, having been found at parts-per-million concentrations. Other pollutants show similar concentrations. Clearly, scientists have justification for their concerns. In fact, we should be concerned about any human-made chemical that routinely occurs (bioaccumulates) in our bodies, since these chemicals tend to be hydrophobic and therefore do not leave our bodies easily. Again, dose is everything. Even though these pollutants may not be acting in a one-to-one cause-and-effect manner as we assumed above, the fact that they, and other chemicals, are in our bodies at the same or greater concentrations than our natural hormones is of grave concern.

As I pointed out earlier, it is also a question of human tolerance and individual response. Remember that humans and other animals do not all respond to a chemical or a particular dose in the same way. As an example, the phytoestrogens (natural plant-derived hormones) that are found in soybeans at the typical concentration found in soy-based foods and in human or animal blood serum are also shown in the bottom right corner of the table. Note the exceedingly high concentrations of phytoestrogens, parts-per-million levels, as compared to the parts-per-billion levels of natural hormones.

It seems, however, that phytoestrogens may not have as pronounced an effect on humans as we might fear, since most women appear to tolerate them well. But recall the diversity of responses discussed above. I have known women who have experienced very dramatic effects, and there are also reports of this in the medical literature. Young women who follow a high-soy diet can have more dramatic menstrual periods. In contrast, older women going through menopause can find considerable relief from symptoms by eating a high-soy diet. Endocrine disruptors can be found in numerous products all around us, and they can interact with our endocrine system in all forms. In males, there is even a report of three pubescent boys developing abnormal breast tissue after exposure to natural lavender and tea tree oils in a skin moisturizer.

Sources of Exposure to Endocrine Disruptors

Today plastics, and endocrine disruptors contained in plastic consumer products, are a hot topic in the media and on the Internet. Table 4.3 provides a list of the most common plastics, by recycling number and plastic polymer. The reason modern society has developed so many different types of plastics is that each plastic has different physical properties. No one plastic effectively serves the needs of every manufacturing and storage process. Some plastics are less expensive; others are better able to withstand heat. Other plastics have additives that give them a desired physical property, such as flexibility. Some scientists have, however, raised concerns over the health effects of certain additives to the plastic formulations. The problem is, we have no definitive toxicological data for many of these chemical additives. Plastics are an important part of our modern life. In some cases, they offer better and safer packing for the products on which we depend. In other cases, they greatly reduce the cost of shipping and handling because of their light weight and nonbreakable nature. But we must choose wisely when using plastics.

Most of our exposure to and intake of known and suspected endocrine disruptors is through our food. These endocrine disruptors can be transmitted to food from even slightly contaminated crop irrigation water,

TABLE 4.3 Common Types of Plastics Used in the Food Industry, Their Uses, and Possible Health Concerns

Number of plastic	Chemical polymer	Polymer abbreviation	Uses	Possible chemical leaching from polymer and biochemical mechanism of concern
#1	Polyethylene terephthalate	PET (PETE)	Soft drink containers	Possible antimony, PET, and phthalates, which are suspected endocrine disruptors
#2	High-density polyethylene	HDPE	Milk crates, milk jugs, beverage bottles, soft margarine containers	None?
#3	Polyvinyl chloride	PVC	Inflatable toys, shampoo bottles, some food containers	Possible leaching of vinyl chloride monomer and phthalate plastic formula additives to food, which are suspected endocrine disruptors
#4	Low-density polyethylene	LDPE	Plastic fibers	None?
#5	Polypropylene	PP	Housewares, screw-on caps, yogurt containers	None?
#6	Polystyrene	PS	Hot food containers, plastic utensils	Possible leaching of styrene monomer and health effects
#7	Polycarbonate	PC (and others)	Many containers for liquids and used in the lining of canned goods	Possible leaching of bisphenol A (BPA) into food, a suspected endocrine disruptor

growing soil, food processing, and certainly food packaging. Central to our concern is persistent concentrations of any unnatural chemical in our bodies. Some of the most common endocrine disruptors, in no particular order, are: (1) the previously discussed dirty dozen POPs that will be with us for decades into the future; (2) chemicals that are used heavily in industrial detergents and as fire retardants (nonylphenols and alkylphenol ethoxylates); (3) any chemical, natural or synthetic, that mimics estrogen activity (xenoestrogens); (4) a component of plastic formulations found in certain medical devices (DEHP, or di-[2-ethylhexyl]phthalate); (5) more chemicals used as a plastic formula agent (represented by a large variety of phthalate-based chemicals); (6) a common flame retardant (polybrominated biphenyl ether, or PDBE); (7) a chemical used in the manufacture of Teflon° (perfluorooctanoic acid, or PFOA); and (8) a variety of hormones exiting from wastewater treatment plants, the water from which is then used in irrigation or taken into drinking-water treatment plants downstream. These chemicals have been found to be accumulating in the environment, in ecological food chains, and in humans. What's next?

Our Next Success(es)

At this point you may be asking, where is the success in this story? First, we have effectively eliminated new inputs of many pesticide-related endocrine disruptors and today attempt to screen for endocrine disruption behavior prior to the mass manufacture of chemicals. And we have finally figured out methods for monitoring, detecting, and predicting what industrial chemicals we need to be concerned with. From time to time you may have seen monitoring programs, targeting workplaces and the general public in order to measure bioaccumulating chemicals in our body (and the environment). One such program, the 1999–2000 National Health and Nutrition Examination Survey (NHANES II), attempted to measure the concentrations of 116 chemicals—from metals to chlorinated hydrocarbons to nicotine—in humans. Once we identify an old or new chemical that is bioaccumulating in the environment or in the human body, it sets off a regulatory alarm for more research. Research is conducted into the source of the chemical(s), what health effects the chemical may pose, and whether the chemical needs more regulation, actual removal from a hazardous chemical list, or banning all together. A significant effort in this regard commenced in the United States around 2000, when Congress mandated that the EPA develop such a program.

Monitoring is the relatively easy part. Predicting, a priori, which chemicals will bioaccumulate and which will be endocrine disruptors is the real challenge. This challenge is being addressed through molecular computational methods borrowed from medicinal drug companies that compare a new chemical's structure to that of known chemicals, natural and synthetic, whose effects on the human body are already known. After a potential problematic chemical is identified by computation, a series of lab tests on animals will follow. Progress is slow, but this approach is better than our historical (and current) approach of using humans, especially infants and children, as lab rats when we introduce the chemical into broad, societal use. Historically, only after we find an adverse effect resulting from a consumer product, will we then regulate or ban the underlying chemical.

One problem that will likely never be solved, however, is the effect of the many combinations of natural and synthetic chemicals to which we are exposed. This is referred to as a synergistic effect—when two or more chemicals combine to result in a more complex reaction or problem. Given the nature of endocrine disruptors and the large number of possible chemical combinations that we are exposed to, this problem will be with us as long as we exist.

Prior to closing this chapter, it is interesting and educational to look at a current poster chemical of the conflict between environmentalists and the chemical industry—bisphenol A, commonly known as BPA. Environmentalists say it is inherently harmful while some in the chemical industry say it is perfectly safe and is a victim of extremist hype. Nonetheless, continued manufacturing as well as the future sales of BPA are bleak.

Bisphenol A was first synthesized in 1891 in Russia, and its estrogen-like qualities were initially noted in the 1930s. Although BPA was not used in estrogen-replacement therapy, similar chemicals were administered up until the 1970s, when they were suspected of causing cancer. But the carcinogenicity of related chemicals has not greatly concerned the chemical industry when it comes to BPA. BPA is used as a component of polymers and as a hardening agent in epoxy resins that are used in food packaging and many other applications. BPA is widely suspected of being linked to reproduction problems and cancer, but industry, with the aid of our government, has fought to continue its use. As evidence of BPA's negative effects on health has increased in recent years, numerous national governments have limited or banned its use, especially in infant products and, more recently, in high human contact products such as polycarbonate-based water bottles (plastics labeled #7 PC). In the United States, the case of BPA illustrates an inconsistency in our country's regulations and their enforcement. While the federal Clean Water and Clean Air Acts have had to rein in less stringent laws in some states, federal action on BPA was surprisingly slow. Individual U.S. states have taken the lead, while the Food and Drug Administration (FDA) seems to be more under the influence of industry and their politicians. Many have voiced concern because of the following:

• Any chemical that bioaccumulates in humans should be scrutinized very carefully, and BPA levels of concern are routinely found in human blood monitoring programs.
• In 2007, a consensus statement by 38 experts on BPA concluded that average levels in humans are above those that cause harm to many animals in laboratory experiments.
• BPA has been proven to have endocrine-like behavior; therefore, it is likely an endocrine disruptor.
• It was also reported in 2007 that of the government-funded BPA experiments on lab animals and on tissues to date, 153 found adverse effects while only 14 did not. Notably, all 13 studies funded by chemical corporations reported no harm from BPA.
• Many, if not all, canning companies are very actively looking for alternatives to BPA-containing chemicals to line their food cans.
• Wikipedia provides a very neutral, lengthy, and informative summary on BPA, part of which is reproduced here: Organizations such as the American Chemistry Council (ACC) have lobbied to make BPA federally recognized as a safe product. The ACC is composed of companies that represent 85% of chemical manufacturing capacity in the United States (Favole 2009). The ACC position states that BPA is safe

for use in products. Government and scientific bodies around the globe have extensively evaluated the weight of scientific evidence on Bisphenol A (BPA) and have declared that BPA is safe as used, including in materials that come into contact with food, such as reusable food-storage containers and linings in metal cans.

Because of consumer pressure, by 2011 BPA was no longer being used in sippy cups and baby bottles. In October 2011, the ACC said that it had petitioned the FDA to make it clear to consumers that BPA is no longer present in those products. Steven Hentges of ACC's Polycarbonate/BPA Global Group said, "What we are trying to do is cut through the confusion and provide some clarity about sippy cups and baby bottles. We want that to be very clear, these products are not on the market. There is no need for parents or consumers to worry about them. They aren't there and they won't be in the future" (Jacobs 2011). In July 2012, the FDA moved to ban BPA in baby bottles and sippy cups to satisfy a request by the ACC, stating that the measure was intended to boost consumer confidence and was not due to safety concerns.

Does this close association between the FDA and a chemical lobbying group seem a bit too close?

But the free market is sending a clear signal to industry. Consumers are siding with academic researchers and BPA is rapidly being taken out of consumer products, certainly products for infants. Product manufactures are running away from BPA. In other cases, however, it is being replaced by even more problematic compounds, such as bisphenol S, an even more potent endocrine-disrupting compound. Where are our government regulators?

The case of BPA is a textbook example of ties between government and industry being too close, as well as of slow movement by some in industry. Citizens are demanding action at the state level while the federal government delays, yielding to political pressure and economic concerns. The declining use of BPA in consumer products is an unexpected, emerging success story: consumer opinions really matter in our market place and they can step in where government protection has failed.

In closing, what cancerous ticking time bombs do we each have in "the very marrow of our bones" or elsewhere? I grew up on a cotton farm and cattle ranch during the 1960s and I was routinely sprayed by a crop duster, most certainly with DDT and who knows what else. Dangers from these persistent chlorinated hydrocarbons have finally been eliminated by citizen environmental movements and scientific research that forced national and international governmental action. Today we are finding that we have whole new classes of suspected endocrine disruptors accumulating in our bodies. What biochemical future and resultant health impacts do these chemicals hold for us? With intelligent efforts and by exercising our power as consumers, can we rapidly eliminate these chemicals from our children's lives and from the environment more broadly? We are on the right track, but we are facing one of the most formidable chemical challenges of human existence.

FIVE

The Safety of Chemicals in Our Food and Water

Risk Assessment

All things are poisons, for there is nothing without poisonous
qualities. It is only the dose which makes the thing a poison.
—PARACELSUS

The Environmental Protection Agency (EPA) deems a safe level of pollution to be emissions that impose a risk of harm to human health no greater than one in one million. My students often rephrase this to ask, "Are you telling me that the EPA and the U.S. government are allowing one in a million people to die every year from industrial pollution? With a U.S. population of 300 million, that's 300 deaths a year!" This latter statement is completely untrue, but it is what the general public think when they initially look at the EPA's approach to human health risk assessment. In fact, as we will see in this chapter, our approaches to human health risk assessment today are so strict that if there was 100% compliance with our pollution and chemical exposure laws, it is *unlikely* that death from legal levels of exposure to poisonous chemicals or carcinogens would occur—unlikely but still possible. Clearly, it is not possible to avoid all chemical exposures in our modern society, but these legal levels reflect what government agencies have deemed safe in terms of cumulative exposure to chemicals. While there can be error in our relatively simple approaches, and risk assessment is premised on probabilities and chance, the United States has established very valid and generally well-accepted methods for human health risk assessment.

In order to understand risk assessment, we must go back a few hundred years to its origins. One of the first great financial risks that could, and that in fact needed, to be assessed, was in merchant shipping. The fundamental question was: If an 1800s-vintage ship was sent to a distant port, what are the chances (risk) that it would be taken by pirates, sink in a storm, hit a reef, or not return for one of many other reasons? Past shipping data was used to empirically predict the chances of each of these disastrous events. Thus, risk assessment was born. This approach was honed over the centuries and has been perfected by one of our largest industries, insurance. You can insure almost anything today—from your car to a dancer's legs—based on the probability of an event occurring. The most common types of insurance are accident and life insurance. This is where our discussion of environmental health risks start—by putting environmental risks in perspective with day-to-day risks that we all take.

Environmental risk assessment is an extremely broad field of study, since the very term *environment* includes everything in our surroundings, including other humans. Of course, here we will narrow our definition mainly to risk associated with industrial and chemical exposure. As with any assessment, if the assessment does not find a detectable risk, then there is likely nothing to worry about and no corrective action will be taken. This is similar to a benefit–cost analysis from business; if the cost of an action, say the building of a manufacturing plant, outweighs the benefit, say revenue generated from product sales, then the manufacturing plant will not be built. In both cases, it does not mean that no risk exists or that no benefit could be garnered from a project, but instead it means that they are too low to warrant changing the status quo.

So, let us first examine the typical causes of death in the United States. The government, and others, collect an amazing amount of data on you each day. Hold on, conspiracy theorists! Most data collection is for relatively benign purposes, from your grocery purchases to health statistics, to determine mortality impacts. These data are collected by the Center for Disease Control and Prevention (CDC). Table 5.1 shows how most U.S. residents died in 2014, the most recent year with available data. The cause of death is listed in the left-hand column and the actual deaths adjusted per 100,000 people is given in the middle. The column on the right scales up to the total number of deaths in the United States, based on a U.S. population of 318,881,992 people in 2014.

First and foremost, a large number of people in the United States die from diseases each year, but an even greater percentage of the population

TABLE 5.1 Common Causes of Death in the United States in 2014

Cause of death	Deaths per 100,000 people	Total number of deaths in the United States
All causes	724.6	2,310,749
Heart disease	167.0	532,563
Cancer	161.2	514,067
Chronic lower respiratory diseases	40.5	129,154
Unintentional diseases	40.5	129,154
Diabetes	20.9	66,650
Motor-vehicle accident	10.8	34,441

in Global South countries die from an even more diverse plethora of diseases. As previously discussed in chapters 1 and 2, a number of diseases still plaguing the Global South have been eradicated in the United States through the provision of clean drinking water and effective disposal of wastewater. It also should be noted that the top four causes of death in the United States—heart disease, cancer, stroke, and chronic lower respiratory diseases—are largely related to tobacco smoke. And herein lies an important distinction: voluntary versus imposed exposure to risk. Overeating and smoking are voluntary choices that help contribute to approximately 1.5 million U.S. deaths per year. There is certainly concern over these deaths, and efforts are ongoing to broadly promote healthy choices. The federal government has also finally held Big Tobacco accountable for the health-care costs related to smoking. But, let us flip the situation. What if an industrial pollution source was found to be responsible for all of these deaths caused by exposing surrounding residents to lethal releases? We would be protesting in the streets, not only in Washington, DC, but also in every state capital and at every industrial site! These two contrasting examples illustrate the difference in attitudes toward voluntary risk (choice) and a risk imposed on an unwilling or unknowing individual by someone else (industry).

Let us look a little deeper at risk. Beyond the 43.8 million smokers in the United States (19% of the population), what other risks do you vol-

TABLE 5.2 Death Rate in the United States for Each Voluntary Action

Action	Annual risk per million people	Total deaths in U.S. based on a 2014 population
Cigarette smoking (1 pack/day)	3,600	1,148,400
Mountaineering	600	191,400
Motor-vehicle accident	240	76,560
Police killed in line of duty	220	70,180
Home accidents	110	35,090
Alcohol (light drinker)	20	6,380
Eating 4 tablespoons of peanut butter per day (resulting in liver cancer due to aflatoxin)	8	2,552

Source: Data (first and second column) are from Wilson and Crouch 1987. Calculations for the third column were based on a 2014 U.S. population of 319 million.

untarily expose yourself to each day? From time to time researchers look at how likely we are to die from actions we take. Intriguing data from Wilson and Crouch (1987), summarized in table 5.2, show how often each voluntary action resulted in death out of one million people at risk. For example, of the 43.8 million people who choose to smoke, how many will die each year as a result of this voluntary action? The answer is 3,600 per million people, meaning that 1,084,103 total people die each year from smoking (based on 2014 populations and assuming that the risk did not change from 1987)! And this does not include the risk to others through second-hand smoke.

As demonstrated by the data shown in table 5.2, humans die in many different ways from everyday activities. But when does a risk of death from another cause, say pollution, become significant? Here we do not mean significant in terms of the value of a life that is lost but instead statistically significant, i.e., the mortality cause by pollution can be isolated from a death resulting from typical everyday activities. This is the key to establishing risk assessment guidelines and to determining fault or no fault by a chemical or industrial process.

The overall death rate in the United States in 2014 was 0.7% (2,231,074 deaths out of a population of 318,881,992 people), or roughly 1%. This will be our starting point. Let us compare this total death rate to some estimated death rates for specific pollution events. As mentioned in chapter 1, treatment of your city drinking water with chlorine in order to kill pathogens can also form carcinogens. Is this something that you should be worried about? Based on robust calculations by the EPA and other experts, it is estimated that the chances of developing cancer and dying from chlorinated drinking water are about one person in the entire U.S. population per year. And here is the problem with trying to remove all risk from our modern society. We cannot statistically, or any other way, tell whether one more person will die each year from drinking chlorinated water, since 2.2 million will already die. Simply put, the signal we are looking for, one person dying, is lost is the background noise (2.2 million dying) of how often we die from other causes. In fact, if we look at data from Global South countries that do not have clean drinking water, we see death rates of around 2,000 per million from diarrheal diseases alone. Another more recent source (Gadgil 2006) puts the childhood death rate at 400 per hour from biological contamination. So, by increasing our cancer rate to approximately 1 per 319 million (the U.S. population) by chlorinating our drinking water, we have simultaneously decreased our pathogen death rate by 2,000 per million. There are many other examples of environmental pollution that result in the same broad conclusion: It is impossible to avoid all risk of death from our modern, industrialized society but we can evaluate each case on a risk–benefit basis and avoid specific chemicals or practices when they are found to be significantly detrimental.

So, what number of deaths should we use as a threshold for changing processes? Given the other voluntary risks we take each year and the many ways that we die, the EPA has decided that a risk of less than one in a million is safe—in other words, that at that low level we cannot conclusively determine that the pollution or emission was responsible for that death because so many other risks are also in play. Before you react to this statement, compare the data in the middle column of the table 5.3: 8,000 per million people will die in the United States each year from mostly voluntary risks that they are knowingly taking, and the EPA will only accept an increase in risk of 0.002 per million before it will take action.

TABLE 5.3 Comparison of Death Rates in the United States

Action	Annual risk per million people	Total deaths based on 2014 U.S. population (319 million)
Overall chances of dying	8,000	2,552,000
Drinking water containing the EPA limit for chloroform (causes cancer)	0.002	~1
Death from drinking microbe-infected water	2,000	638,000

Or to put this in real numbers for the total U.S. population, 2.6 million deaths will occur before we can detect the 1 death caused by chlorinated drinking water; yet even then 2,000 people would die from a disease caused from not drinking chlorinated water. This approach seems to strike a good balance between providing a safe environment and not completely shutting down modern society or forestalling the health benefits from new technology and innovation that it can provide.

Next, we need to understand how the EPA determines safe levels of chemical concentrations. The EPA defines *safe* both from the obvious standpoint of not causing death to humans but also as not causing any detrimental health effects. There are three approaches, all of which must be pursued before a chemical can be deemed safe: (1) determining whether a chemical is an endocrine disruptor and at what concentration (this was discussed in the chapter 4); (2) determining whether a chemical will cause cancer and at what concentrations; and (3) determining whether a non–cancer-causing chemical will cause other detrimental health effects or death.

Predicting Cancer Risk from Chemical Exposures

Methods for estimating whether a chemical will cause cancer are highly disputed, but they all basically involve developing a dose–cancer rate response line such as the one shown in figure 5.1. This line is usually based

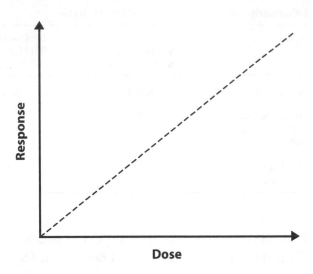

FIGURE 5.1. Cancer formed as a function of pollutant dose rates for a cancer-causing agent

on experimental data from lab animals, but some human data are available from historical pollution events.

Once this dose–response relationship is established, it enables researchers to predict the expected rate of cancer based on a given dose (daily intake dose of a chemical). We now have fairly simple equations for predicting cancer rates from a variety of exposures, including obvious ones such as drinking polluted water, breathing polluted air, or eating contaminated food, as well as less obvious but still important exposure pathways such as dermal contact with pollutants while swimming, children eating polluted soil (in their tasty mud pies), and adsorption of pollutants from skin contact with polluted dirt.

There are, however, important assumptions that underlie the seeming simplicity of these calculations. In particular, there are standards for integral variables that have to be agreed upon by the EPA and industry, such as how much water does an average person drink per day (2.0 liters, about one-half of a gallon), how much air one breathes in a day (20 cubic meters), an average person's weight (70 kg), and average life expectancy (70 years). There are allowances or differences between calculations for children as compared with adults, although recently there have been efforts to further distinguish between children and infants.

After these values are agreed upon and the dose–cancer rate is established for a given pollutant, calculations are performed; if the risk is greater than one instance of cancer per million people, the risk is deemed to be too great. But note that here we are only considering cancer formation, not death, so the level of protection is actually even greater than previously discussed. As mentioned earlier, based on the many other factors that can cause death, we will not be able to detect any increased risk to humans if we limit the risk to one in a million. Most environmental scientists are very comfortable with this approach.

Predicting Noncancerous Health Risks from Chemical Exposures

Risk determination for non–cancer-causing chemicals is harder to assess and certainly harder to explain from a biochemical toxicological perspective. For one thing, in the above risk assessment of cancer-causing chemicals, there is no dose that is considered to not cause cancer, which is why the left end of the dose line in figure 5.1 goes to zero risk at zero concentration, but there is an acceptable incidence of cancer, one in a million. Risk assessments of non–cancer-causing chemicals also rely on dose–response curves from lab animals or historic pollution events involving humans. We again arrive at a dose–response curve, but unlike with carcinogens, these curves are not straight lines starting at zero response for zero concentration. In other words there are, or appear to be, safe exposure doses for some of these chemicals.

In order to economically test these chemicals we must expose animals to pollutant concentrations that are sometimes higher than those to which humans would be exposed. Even with the greatest diligence toward reducing costs and streamlining testing, a simple one-generation test can involve 600 animals, usually rats, and take weeks to months to complete. Using lower chemical concentrations would require even more lab animals, much longer time frames, and considerable, even prohibitive, expenses. In order to avoid this expense, we rely on some assumptions. A generic dose–response line or curve is shown in figure 5.2. Note that most

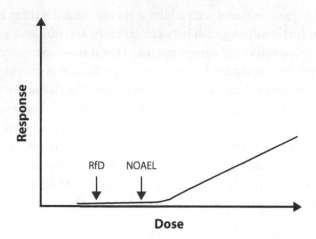

FIGURE 5.2. Health effect as a function of dose for a non-cancer-causing pollutant (NOAEL=no observed adverse effect level; RfD=reference dose)

of the curve is linear but it becomes nonlinear at low doses or exposure. There are several notable calculated points along the curve but here we will discuss only the point corresponding to no observed adverse effect level (NOAEL) This is the pollutant dose below which researchers see absolutely no harmful animal response. However, because there are variations between animal and human responses to a chemical, when no human data are available and animal data must instead be extrapolated to predict human responses, the NOAEL is divided by 1,000 to account for uncertainties and produce a sufficiently conservative risk number. The result is the RfD, or reference dose, a dose that researchers feel is perfectly safe for human exposure. This approach builds in considerable safety, since it results in levels of allowed exposure that are 1,000 times lower than any observed effect. This seems like a very reasonable and safe approach, since we will never be able, and it may in fact not be optimal, to avoid all chemicals nowadays.

A good question to examine at this point is: What triggers a risk assessment test or study? We certainly cannot afford to do such a study on every chemical that has been synthesized. There are many reasons why the EPA may require or conduct a risk assessment study. There may be a high incidence of cancer observed in a particular population that warrants investigation. Or a contaminated piece of land or water system may be under review for designation as a Superfund site. Or a new chemical may be about to enter the market that, based on previous chemical knowledge, is expected to have a high toxicity. The EPA has been mandated by Congress to establish a very careful plan for risk determination in each of these instances. The first step is hazard identification, which, as suggested by its name, investigates what problem or health effect may be caused by which chemicals. Next, experiments are conducted to construct a dose–response curve or line. Then, an exposure assessment study examines the anticipated frequency, timing, and levels of contact between humans and the chemicals. This is where very straightforward exposure equations are used to estimate actual risk. And finally, a risk characterization is completed to examine how well the data support proposed policy solutions and regulations governing the chemical. This process has been completed hundreds if not thousands of times since the implementation of risk assessment, but given the large numbers of chemicals, as well as the new chemicals entering the market, the EPA is substantially backlogged on toxicity determinations and approvals of chemicals.

So far we have only discussed human health, but the EPA's clear mission is to "protect human health and the environment" (although several administrations have tried to avoid the "environment" part of the mission). What about the risks chemicals may pose to other animals or to ecosystems? Since the 1990s, the EPA has developed highly sophisticated methods for animal and ecological risk assessment. The steps required for ecological risk assessment is similar to those for human health assessment. First, an environmental hazard is identified or a chemical of concern is proposed to be put on the market. Second, animal-specific toxicity testing is conducted. Next, chemical and physical data such as water solubility, volatility, and many other parameters are measured to determine how the chemical will act in the environment. For example, this step asks, will the chemical mostly be found in the water or in sediment and, will the chemical readily biodegrade or chemically degrade? Finally, more testing is conducted concerning magnitudes of exposure. This testing protocol is essential for old and new pesticides and for particularly toxic chemicals. If these methods had been in place in the 1940s and 1950s, we probably would have identified the harms of dichlorodiphenyltrichloroethane

(DDT) discussed in chapter 4 and thereby avoided the near extinction of the American bald eagle and other apex birds of prey. Today, we have effective risk-assessment procedures in place to avoid similar disasters in the future.

Next Steps in Risk Assessment

There are still problems with our approach to risk assessment. The first is that risk assessment produces a single metric that people often myopically promote while ignoring other factors that also need to be figured into a thorough assessment. In fact, this was the basis of a semi-conspiracy theory perceived by thoughtful environmentalists in the 1980s when risk assessment was pushed during the Reagan administration. The fear, that has to some extent come true today, was that if a single, supposedly comprehensive number is generated, it will be used in preference to other factors. I have been active in two Superfund sites and have observed many other Superfund Remedial Investigations—Feasibility Studies. Unfortunately, in my experience, the biggest factor in decision making *has* been the risk-assessment number(s), while other factors, such as sensitive geographical locations or populations and ecological (versus human) concerns, have weighed less in the decision process. This should be a concern in future assessment efforts.

A second, and major, limitation of risk assessment is that most dose–response curves evaluate only one pollutant at a time. Rarely does a polluted site only contain one pollutant. Take something as seemingly simple as the water runoff from a grocery store parking lot after a rain storm; a chemist can find components of gasoline, polyaromatic hydrocarbons, inorganic and organic components from automotive tires, chemicals from tobacco, and many other compounds in the water, not to mention fertilizers and herbicides from nearby lawns. For a more complex, industrial site, the number of chemicals contained in an emission plume greatly increases. Failure to capture the combined effect of chemicals contained in a single exposure presents two problems. First, what is most important in determining risk is the aggregate effect of exposure. Second, the combination of chemicals can interact with one another to have different toxicities and thus present different risks. Therefore, when there is a human or ecological problem, we must have the ability to assess combinations of chemical exposures.

The third limitation of risk assessment is completely political. As we will discuss in chapter 7 on environmental law, the person or political party in power in a state or within the federal government greatly influences the direction and aims of environmental policy. This can include prioritization of which polluted sites are assessed and which factors are given the most weight in final remediation decisions. While human health risk assessment can be an important driver in this process, the cost of remediation can be determinative. Potentially most disturbing, the EPA has actually put a value on human life for use in benefit–cost estimates. The current value of a human life, in terms of "pricing" an environmental cause of death, is around $7.8 million. (Needless to say, other species are valued at a much lower number.) So, how is this number used? As noted by Seth Borenstein, an Associated Press science writer, "Consider for example, a hypothetical regulation that costs $18 billion to enforce but will prevent 2,500 deaths. At $7.9 million per person, the lifesaving benefits outweigh the costs." (Borenstein 2008). But what if the numbers turn out differently? What dollar value would you put on your life? Or your child's life?

Overall we have fairly good procedures in place in the United States and many other countries to assess human and ecological exposure to natural and human-made chemicals, but as always, our system can improve. Research scientists are highly committed to improving our techniques, and while some of my fellow environmentalists will take issue with me calling risk assessment a success story, we are off to a good start. Perhaps the most important next step is to explain our current risk-assessment approaches to the general public in a clearer, more effective way. I hope this chapter helps in that effort.

SIX

Saving Our Atmosphere for Our Children

The new American finds his challenge and his love in the
traffic-choked streets, skies nested in smog, choking with the
acids of industry, the screech of rubber and houses leashed in
against one another while the town lets wither a time and die.
—JOHN STEINBECK

Saying sulfates do not cause acid rain is the same as saying
that smoking does not cause lung cancer.
—DREW LEWIS

The hole in the ozone layer is a kind of skywriting. At first
it seemed to spell out our continuing complacency before
a witch's brew of deadly perils. But perhaps it really tells
of a newfound talent to work together to protect the
global environment.
—CARL SAGAN

All of the above quotes concern components of our Earth's atmosphere. The first is highly geographically localized and refers to urban smog. The second refers to acid rain, a more regional problem we completely understand, everyone agrees on, and that we have addressed in the construction of new installations, and we could easily modify old installations to avoid this problem if our government had the willpower. The third deals with our Earth's stratospheric ozone hole, an issue that is finally understood and that we are making every possible effort to remediate. This chapter is about success in saving our atmosphere for life as we know it. But, before we delve into atmospheric changes, first we need to go back to the beginning, the very beginning of our planet's history.

Our solar system emerged about 4.6 billion years ago, long after the Big Bang that occurred 13.8 billion years ago. The time just after our solar system formed was a violent one. The Earth is thought to have initially been composed of a very reducing atmosphere, lacking our precious oxygen, but abundant in hydrogen (H_2), water vapor (H_2O), methane (CH_4), and a little ammonia (NH_3). This is commonly referred to as our first atmosphere. But young stars, such as our Sun 4 billion years ago, have very strong solar winds, and most, if not all, of our early atmosphere was literally blown off the planet and into space. After the initial 1 billion years or so of violent events, including the collision of large asteroids with Earth, tectonic events on our planet released considerable volumes of gases such as water vapor (H_2O), nitrogen (N_2), and carbon dioxide (CO_2) that formed our second atmosphere. At this time, a lack of a strong solar wind and far fewer large-asteroid collisions produced a relatively stable period starting around 3.3 billion to 3.8 billion years ago. This more stable environment allowed the first life forms to develop on Earth.

The first photosynthetic organisms are thought to have been cyanobacteria, or blue-green algae. These very primitive algae, which are still abundant today, took carbon dioxide (CO_2) from the "second" atmosphere and coupled it with water to form the first diatomic oxygen (O_2) on Earth. This oxygen was essentially a waste product for blue-green algae and was toxic to many other bacteria living at that time. Populations of blue-green algae grew and pumped out more and more oxygen over the next billion years. By approximately 2.2 billion to 2.5 billion years ago, atmospheric oxygen levels reached 1% of the composition of the atmosphere, as methane (CH_4) and carbon dioxide (CO_2) levels had greatly decreased. I mention this because today many skeptics of human-caused global warming, and previous skeptics of acid rain and ozone depletion, question how human activities could cause a change in something as vast as the atmosphere. The answer is simple. If a nonthinking, prehistoric, single-celled organism could alter the atmosphere billions of years ago, humans can certainly do so today, knowingly, and in many different ways. We'll discuss this more later in this chapter and in chapter 9 on climate change.

The beginning of oxygen accumulating in the atmosphere represented the start of Earth's third atmosphere. Oxygen levels continued to rise and allowed more advanced forms of life to evolve as far back as 1.3

billion years ago. Approximately 1 billion years ago, oxygen levels reached approximately 10% of the atmosphere and they increased further, to 15%, 0.54 billion years ago. Levels have fluctuated in the past 0.5 billion years from 30% to 21% today. Oxygen has now steadily comprised approximately 21% of the atmosphere for the past 150 million years. This 150-million-year-old balance between oxygen production and consumption has given Earth a relatively stable atmosphere for advanced life to evolve into today's global ecosystem based on the natural functioning of microbes. Imagine what a sentient, apex human species can do to the atmosphere today with all of our industrial resources!

Smog

Many locations around the world are known for fog and stagnant atmospheric conditions, but when you add selected industrial emissions and home coal-burning releases to the mix you create a hazardous, ever-more-deadly, concoction of airborne pollutants. Such were the circumstances in London in 1952 when smog settled on the city for five days and is blamed for thousands of premature deaths. In the United States, Los Angeles suffered from stagnant air and atmospheric releases from industry and automobiles to produce some of the dirtiest air in the world in the 1940s. And yet governments around the world learned little from these deadly historic events, as citizens of Asia, especially in Beijing, China, suffer today from the same set of avoidable pollution events.

But what is smog, and how is it formed? Today smog, more correctly referred to as photochemical smog, is the result of sunlight reacting with the human-made atmospheric releases of volatile hydrocarbons and nitrogen oxides to produce a hazardous, and even toxic, mix of ozone and nitrogen dioxide (NO_2) that harms our lungs. Volatile hydrocarbons come from organic releases from automobile fueling operations, chemical industries, and petroleum refineries. Nitrogen oxide (NO) is produced by all high-temperature processes; the most notable in terms of our discussions are coal-fired power plants and automobiles. Nitrogen oxide further reacts in the atmosphere to produce NO_2, the precursor to ozone and smog production. The most effective way to eliminate ozone production is to remove one of the major precursors. Of course, removing sunlight is not an option, although it should be noted that smog production does not occur during nighttime hours even in the most polluted cities. Volatile hydrocarbons are everywhere in highly populated areas, but considerable advances have been made to reduce emissions from automobile fueling operations by installing vapor-recovery systems on gasoline pump nozzles in California and flares at petroleum refineries. But the control of NO emissions was a problem given the number of automobiles in these highly populated areas.

Technology was the solution in the form of catalytic converters that converted the NO produced in automobile combustion engines back to harmless atmospheric N_2 gas. But as noted in chapter 3, the catalytic converter was poisoned and rendered useless when leaded gasoline was burned in the car. Again, this resulted in a win–win situation for humans and the environment. The required factory installation of catalytic converters required the production and use of lead-free gasoline, and these new automobiles greatly reduced NO emissions. And as noted in chapter 3, the cost of pollution abatement was directly transferred to the consumer when purchasing a new car. Industry has also been required to remove NO from many of their atmospheric emissions.

Air quality, in the form of smog reduction, has greatly improved in a number of Global North cities, most notably in Los Angeles and metropolitan areas on the Eastern Coast. But this improvement did not occur overnight. After the required installation of catalytic converters and the removal of lead in gasoline in the late 1970s, lead pollution levels dropped relatively quickly, and by 95% in 1990. Ozone, NO_2, and other automobile emissions dropped more slowly because of the continued use and slow retirement of old vehicles; but these emissions have dropped considerably today. While residents of the Global South are mostly free of leaded gasoline today, they are still lacking in the inclusion of catalytic converters in their various modes of transportation. Thus, they still suffer from smog events in selected areas.

Acid Rain

In modern times, our second, reasonably effective, success story in addressing atmospheric pollution was controlling or limiting acid rain (acidity in precipitation). With the exception of when we fly in an airplane or spacecraft, humans reside in the tropospheric portion of the atmosphere, which contains 80% of the mass of the atmosphere and ranges from a height of 55,000 ft (17 km) at the equator to a height of 30,000 ft (9 km) at the poles. This is where we live, breathe, and release all of our gaseous pollutants. Although volcanoes release gases that can form acid rain, the first human-produced precursor of acid rain was from burning fossil fuel, mainly coal. There are two main chemical culprits: NO and sulfur dioxide (SO_2). These compounds are more commonly referred to as NO_x, since NO and NO_2 form nitric acid (HNO_3), and SO_x, since sulfur dioxide (SO_2) and sulfur trioxide (SO_3) form sulfuric acid (H_2SO_4).

Before discussing the solutions we have begun to implement, we need to examine the chemistry of acid rain formation that no one, including the coal industry, now disputes. Acid, of course, refers to the pH scale that most readers will be familiar with from swimming pools. The scale, ranging from 0 to 14, is logarithmic and is related to the solution's hydrogen ion concentration. Perfectly neutral water has a pH of 7.0. Surprisingly to most people is the fact that natural water, including rainwater, has a pH value lower than 7.0. This is due to the presence of carbon dioxide (CO_2) in the atmospheric, which dissolves in water and subsequently reacts with it to form very low concentrations of carbonic acid (H_2CO_3), the same acid found in carbonated spring water and other beverages at a much higher concentration. Given today's CO_2 concentration in the atmosphere of 380 to 400 ppm, the pH of rainwater is about 5.5. So, by definition, any water with a pH less than 5.5 is considered to be acid rain.

In the atmosphere, nitrogen is mostly present as the diatomic, triple-bonded and very stable molecule, N_2, which is in equilibrium with diatomic oxygen, O_2. The two molecules do not normally react with one another to any extent. But, when the temperature around these molecules is greatly increased, N_2 and O_2 react to form nitric oxide (NO). This temperature increase can occur naturally during a volcanic event or in the presence of lightning, or unnaturally through a high-temperature combustion event such as in an automobile engine or a fossil fuel–fired furnace at a natural gas, petroleum, or coal-fired power plant. Chemically, NO is known to be a free radical, meaning it is very reactive. In the atmosphere it will react with another free radical, the naturally occurring hydroxide radical (OH·) that is formed by the ionization of water vapor by ultraviolet light. The resulting peroxy radical ($HO_2\cdot$) then converts nitric oxide to nitrogen dioxide (NO_2), another free radical compound. NO_2 in turn reacts with water vapor to form HNO_3, a strong and corrosive acid.

Sources of SO_x are considerably different. Sulfur is a natural component in coal and other fossil fuels because it was present in the amino acids of the now-long-dead plants that comprise fossil fuels. Sulfur was also present in the ancient environment when coal was formed; it reacted to produce iron sulfide during the fossilization process. Thus, all forms of sulfur in fossil fuels are reduced, meaning that they have an abundance of electrons and no oxygen atoms are bound to them. When SO_x and NO_x are oxidized to sulfuric acid and nitric acid they form strong acids that create acid rain. The amount of sulfur contained in coal varies with the source of the coal, ranging from up to 3% by weight in lignite coal to less than 1% in anthracite coal. Today the problem is one of scale. China is the largest producer of coal, producing 3520 million tons in 2011, while the United States has the greatest reserves, estimated at 237,000 million tons, or 22% of the world's supply. Untreated coal combustion at power plants is the greatest source of SO_x and therefore the greatest concern with regard to acid rain. During the combustion process, an oxidation reaction takes place. Specifically, sulfur undergoes a chemical reaction to form SO_2 and then further reacts in the atmosphere with the hydroxide radical mentioned above to form SO_3. SO_3 dissolves in water and forms sulfuric acid, H_2SO_4, another strong and destructive acid.

We call both HNO_3 and H_2SO_4 strong acids because they completely dissociate when they come into contact with water in the atmosphere and donate their hydrogen ions as acidic ions (known chemically as H^+). Acid rain harms plants as well as some human-made materials as it falls. Some soils have natural components, such as limestone (calcium and magnesium carbonate, $CaCO_3$ and $MgCO_3$), that can neutralize these strong acids. A problem arises, however, when only small amounts or no natural carbonates are present in the soil. When this happens, the strong acids in acid rain lower the pH of the soil. This in turn releases naturally occurring, and normally benign, aluminum from soil minerals in the form of an aluminum ion (Al^{3+}). The aluminum ion is toxic to most aquatic organisms as well as to plants growing in the soil. While acid rain can be harmful on its own, when it comes into direct contact with plants and various human-made materials, the dissolved aluminum ion (Al^{3+}) is the true toxic result of acid rain.

Natural sources of acid rain have been present on Earth since before human habitation but generally only in small amounts. One exception is major volcanic eruptions, which can and have devastated large areas of land. The first fires made by hominids and forest fires caused by lightning also produced small amounts of acid rain, but Earth's natural neutralization processes were reasonably able to manage these rainfalls. The phenomenon of acid rain was first noted in the mid-nineteenth century, and the term was coined by Robert Smith, an English scientist, in 1872. Although not a problem initially, by the 1970s, acid rain had become widespread because of the increasing use of coal to generate electricity. In an effort to assess the problem, Congress passed the Acid Deposition Act in 1980, a 10-year research program created to study the impact of acid rain on freshwater and terrestrial ecosystems, weathering of historical buildings, and damage to contemporary building materials. The first findings reported in 1991, found that 5% of New England lakes were acidic. Also in 1991, Congress passed a series of amendments to the original Clean Air Act that addressed NO_x and SO_x emissions. In particular, the amendments to the Act called for a decrease in SO_x emissions from power plants of 10 million tons that was to be achieved in two phases using a cap and trade program. Between 1990 and 2000, monitoring stations in New England revealed a 25% decrease in NO_x and a 54% reduction in SO_x emissions. In 2005, the Environmental Protection Agency (EPA) passed the Clean Air Interstate Rule, which, when fully implemented, will reduce SO_x emissions in 28 eastern states by 70%.

The Clean Air Interstate Rule points out one major problem with NO_x and SO_x emissions; most emissions come from the coal-fired plants located in the midwestern United States. As weather patterns push air masses that form and contain acid rain to the northeast, however, acid rain falls out over New England and Eastern Canada. The first problem requiring a remedy was the lack of interstate and international rules concerning the generation of acid-rain precursors. The second was one of geology. Most soils in this eastern region of North America have little to no pH buffering capacity, since they are deficient in carbonate minerals. This is a major concern for the beautiful lakes in the New York Adirondacks, as they have been largely acidified.

In the United States, the biggest hurdle to acid-rain legislation was the lack of emissions control on old, coal-fired power plants. This concept, known as a "grandfather clause," allowed older plants to not update their pollution-control equipment since it was anticipated that they would soon be shut down because of their age. Not surprisingly, many of these plants continued operating past their proposed shutdown date with no additional pollution-control equipment. In an attempt to address this, acid-rain legislation firmly called for the installation of SO_x and NO_x pollution-control equipment with best available technology if any major modification to a power plant was undertaken. Many of these older plants did in fact undergo major modifications, but unfortunately, their operators refused to install pollution-abatement equipment since it would create added costs and they claimed that the modifications were only maintenance. The EPA and other environmental groups took these violators to court. They were winning these enforcement lawsuits until the George W. Bush administration stepped in and, in typical industry-supported fashion, decided to stall further lawsuits, their argument being that the problem needed to be studied more. Studying the problem was a well-practiced Reagan administration approach to delay any action as long as possible. Unfortunately, George W. Bush did not follow his father's very effective approach to acid-rain abatement. George W.'s tactic was used particularly voraciously with respect to global warming, as will be discussed in chapter 9.

Today we are back on track with acid-rain remediation in the Global North. Few, if any, coal-fired plants are being constructed, many coal-fired power plants are being retrofitted with SO_x and NO_x pollution-control equipment, and older plants will soon be finally shut down or retrofitted. In addition, because of railroad deregulation, cleaner, lower-sulfur coal (not to be confused with nonexistent clean coal) from the Powder River Basin in Wyoming is now being more commonly used across the nation, especially in the Midwest. Because of weather patterns and the rapid-formation chemistry of acid rain, it is a largely U.S. domestic phenomenon. So Global North countries are, or soon will be, in very good standing with regard to solving this problem. China, on the other hand, is a completely different story. With a coal-fired power plant being constructed about once a week, which then burns China's relatively high-sulfur coal, the region is just beginning to experience decades of acid-rain damage. It is time again for China's citizens to stand up and demand action. Studies in Global North countries have found that the relatively small costs of acid-rain pollution abatement are well worth the social, economic, and environmental benefits.

The recurring theme in pollution abatement is unnecessary delay as a result of a few politicians whose strings are controlled by the fossil fuel or chemical industry. These industries, and their political puppets, know that change will eventually occur, but the longer they can delay the change the more profits they can make at the expense of nature and human lives. A relevant example here is the delayed but successful cap and trade program for reducing sulfur emissions from coal-fired power plants. This led directly to technology innovations and lower-than-expected costs of reduced sulfur emissions. Will we ever learn from our history with a few bad actors from industry, or are we destined to repeat it over and over— e.g., with the global warming argument? To summarize, every year implementation of a law or technology is delayed means more profits for bad actors in industry and more costs to human health and the environment.

Ozone Depletion

For our next largely successful atmospheric story, we must travel up from the troposphere that we live in to the stratosphere, where a protective layer of ozone resides. Our stratospheric ozone layer started to form slightly less than a billion years ago when tropospheric oxygen levels increased to 10% from the growth of cyanobacteria and their excretion of oxygen. Ozone forms high in the atmosphere because of incoming relatively high-energy ultraviolet (UV) radiation from our Sun. This radiation is so energetic that it breaks the diatomic oxygen (O_2) that absorbs it into single, but very reactive, radical oxygen atoms. Single oxygen atoms are not stable in their radical form and quickly react with nearby atoms, which happen to be other O_2 molecules. This reaction forms the relatively stable, but also radical ozone molecule, O_3.

The absorption of UV radiation by Earth's atmosphere during the ozone formation process is critical because otherwise this high-energy radiation would harm life on Earth's surface. In addition, ozone, once formed, also absorbs UV radiation. Therefore, the formation of our ozone layer a billion years ago was a vital step in the development of the Earth's atmosphere that removed life-harming radiation from the Earth's surface and allowed our single-celled microbial ancestors to crawl out from under their rocks and start the evolutionary process that eventually led to higher life forms. A few hundreds of million years later, the human species evolved that would come to appreciate the ozone layer, but only after nearly destroying it.

Now for the story of how we, completely unknowingly, nearly destroyed our protective blanket of ozone. The main function of the ozone layer from our perspective is to filter out harmful UV radiation coming in from our Sun. As any beach-goer knows, a little UV radiation still reaches the surface but far less than would be present if our ozone layer was not up in the stratosphere. It is also important to distinguish between stratospheric ozone and the ground-level ozone with which you may be familiar. In contrast to our protective ozone layer, ground-level ozone causes harmful air pollution such as smog.

The story of stratospheric ozone depletion starts back with the mass production and use of refrigeration units. Our modern, vapor-compression cycle refrigeration was first developed in the early 1800s. Until the 1920s, these units used relatively reactive and toxic chemicals, such as ammonia and sulfur dioxide, as the refrigerants; these substances were not entirely safe in the home. General Motors recognized the need for a safe, nontoxic, and nonflammable alternative. To solve this problem, they went to their proven chemist, Thomas Midgley Jr., of tetraethyl lead fame who you may recall from chapter 3, and Albert Henne. The team successfully developed the first class of compounds known as Freons, specifically dichlorofluoromethane, known today as Freon-12. Unlike the harms posed by tetraethyl lead, which were known from its first synthesis to its demise in the 1970s, initially there were no real concerns expressed about Freons. All practical and scientific knowledge about these compounds indicated that they were perfectly safe because of carbon's strong and non-biodegradable bond with fluoride and chloride. Simply put, no adverse health or environmental effects were noted or suspected. The development of Freons was therefore heralded as a great technological advancement for commercialization of industrial and home refrigeration units and it greatly improved our ability to safely store food and reduce food waste.

However, over the decades, Freons—known technically as chlorofluorocarbons (CFCs)—accumulated and rose up into the tropopause, a layer of the atmosphere above the troposphere where we live. Eventually Freons will pass in very small parts-per-trillion concentrations from the tropopause into the stratosphere, where they do degrade. This process is again based on ozone chemistry. Because of the ozone formation reaction in the stratosphere discussed earlier, very little UV radiation reaches the troposphere. Instead, essentially only visible and infrared (IR) radiation are present in the troposphere and these wavelengths of radiation are too weak to break the carbon–chlorine bond in the CFCs. But, as CFCs rise into the stratosphere, where high-energy UV radiation is present, the carbon–chlorine bonds are broken and the reaction produces radical chlorine atoms. This reaction occurred in the stratosphere for decades before scientists discovered it.

As is often the case with science, we first noticed the effect of stratospheric ozone depletion, or ozone-hole formation, and only after the fact found the cause, CFCs. Ozone depletion came to fame when a single scientific publication in 1985 (Farman et al. 1985) reported stratospheric ozone depletion using ground-based measurements. In addition, NASA launched many satellites into Earth's orbit in the 1960s but few came to fame. One of these was the total ozone mapping spectrometer (TOMS) on *Nimbus 7* that was launched in October 1978 to monitor stratospheric ozone concentrations over Antarctica; TOMS confirmed the ground-based measurements. Significant decreases in ozone concentration were noted by TOMS instruments both back then and still today. If you do an Internet search on "ozone hole map," you can view many color-ozone images that capture this phenomenon and show the decreasing growth rate of the ozone hole since we discovered and started to address this problem.

Shortly after reports of the scientifically documented phenomenon were published, the lay media sensationalized our impending global doom. But as mentioned above, when we first recognized this depletion, we did not know what was causing it. We knew that it mostly occurred above the South Pole as well as to a lesser degree above the North Pole, but why? CFCs were known to be accumulating in the stratosphere, but the exact chemical reactions were not known. People began to ask in the 1970s, should we immediately ban CFCs?

As early as 1974, Frank Rowland and Mario Molina, chemists at the University of California, proposed that CFCs could act in a destructive manner in the stratosphere in causing ozone depletion. The idea that chlorine radicals could be formed in the stratosphere and, in theory, react to destroy ozone was highly disputed by the main manufacturer of CFCs, DuPont, in the typical "not my chemical" reaction of industry. DuPont referred to the CFC link to ozone depletion as a "science fiction tale . . . utter nonsense" (Masters 2016). You may note the similarity to industry's response to climate change, but we will get to that subject later. Denial continued for several years, but as mounting evidence documented the presence of chlorine radicals and chlorine oxide radicals in the stratosphere and as proposed chemical mechanisms of ozone depletion were backed by laboratory experiments, industry's cries fell more and more on deaf ears. Two facts put the nail in the coffin for CFCs. The cause-and-effect plot, reproduced in figure 6.1, showed increasing chlorine monoxide radical concentration and decreasing ozone concentration. These data, measured during a flight high over Antarctica, clearly shows that stratospheric ozone concentrations plummet in the presence of OCl.

The second piece of evidence was when Rowland and Molina finally proved the chemistry in the lab. The actual chemical mechanism behind the destruction of ozone may be one of the most complicated chemical mechanisms in stratospheric chemistry. The missing link was the formation of catalytic nitric acid ice crystals on very cold, stratospheric clouds in early spring. Rowland and Molina's efforts earned them the 1995 Nobel Prize in Chemistry. Not a bad reward for a "science fiction tale . . . utter nonsense"!

But what could be done about chemicals that had been emitted into the atmosphere for decades and that can reside there for 50 to 100 years? How much ozone can we lose and still maintain life on Earth? Was life on Earth, in fact, doomed? Clearly the first step was to phase out and eventually ban CFCs, much to DuPont's chagrin. Thankfully, governments acted quickly once the science of ozone depletion emerged. It is worth pointing out the difference between smog and acid rain (largely local phenomena) and ozone depletion (an international issue of the atmospheric commons). In ozone depletion, the global community managed to cooperate largely on the creation of a cost-effective substitute, in contrast

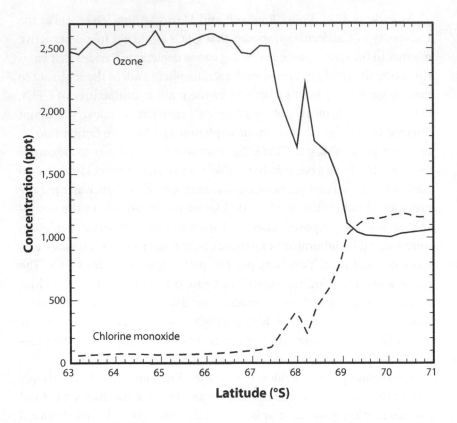

FIGURE 6.1. Atmospheric ozone and chlorine monoxide concentration (ppt=parts per trillion). Data from the NOAA Earth System Research Laboratory (esrl.noaa.gov).

to the failure of the United Nation (UN) to pass binding, enforceable commitments on climate change, the subject of chapter 9.

The UN passed the Vienna Convention for the Protection of the Ozone Layer in 1985. In 1989, the Montreal Protocol on Substances that Deplete the Ozone Layer was signed; this set a timetable for the phaseout of most CFCs. And then in 1995, Rowland and Molina earned the Nobel Prize for their work on stratospheric ozone depletion. The Montreal Protocol has worked very well, as can be seen in figure 6.2, which shows decreases in annual CFC production and replacement by hydrofluorocarbons (HFCs); the documented decrease in stratospheric CFCs from 1980 to 2006 is shown. This was one of the most successful moments of the UN.

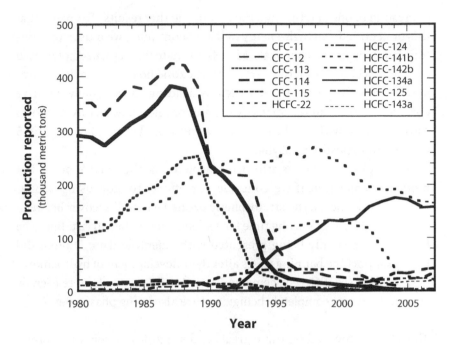

FIGURE 6.2. Atmospheric CFCs and HFCs. Data from European FlouroCarbons Technical Committee (http://www.afeas.org/data.php).

One major problem with the CFC phaseout was the still urgent need for inexpensive refrigeration in Global South countries, especially highly populated Asia. While only 25% of the world's population lived in the Global North at the time, they used approximately 80% of manufactured CFCs. Asia desperately needed their refrigeration industry to scale up to Global North levels in order to feed their growing and urbanizing population. The participation of Asia in the Montreal Protocol was successfully negotiated using technology transfer between Global North and South countries.

So, have we solved the problem of ozone depletion? Will life on Earth survive? The general consensus is yes, even given the generally accepted fact that each CFC molecule will stay in the atmosphere for 50 to 100 years and will destroy approximately 200 or more molecules of ozone during that time. Remember that new ozone is created every day as incoming radiation from our Sun reacts with O_2 in the stratosphere. As previously discussed, harmful UV radiation is absorbed both through the process of creating ozone and through the new ozone that results. To ensure that this protection from high-energy radiation continues, we must stop emitting ozone-depleting chemicals. Models estimate that if we had not phased out CFCs when we did, the stratospheric ozone hole would have continued to grow far past the polar regions, leading to extremely negative consequences for all life forms. As a point of reference, NASA scientists estimated that without the ozone layer, you would be sunburned after 5 minutes of exposure to daylight.

Stratospheric ozone is still being depleted at the Earth's poles, but models predict that if we continue our efforts to stop using ozone-depleting chemicals, the stratospheric ozone layer will recover between 2050 and 2065. That's a long time for a science fiction tale to last. But to be fair, after repeatedly being presented with scientific data, DuPont did finally support the banning CFCs after their development of hydrochlorofluorocarbon (HCFC) replacements. Today HCFCs, a much more friendly chemical, are not completely benign and are also being phased out.

This chapter focused on three great success stories for what was previously thought of as an inexhaustible resource, our atmosphere. Remediation or minimization of localized smog is rapidly becoming a success story in the Global North. Acid rain is a very controllable problem in Global North countries but now urgent action is needed in Asia, especially in India and China. The technology is simple and effective for this regional pollution problem. Likewise stratospheric ozone depletion has been addressed. While harmful chemical emissions have stopped, we still have to study the recovery progress for decades and be on the lookout for similarly harmful chemicals. But all is not well with our atmosphere. We will return to this topic in chapter 9 on climate change—humans and the Earth's greatest challenge.

SEVEN

Legislating Industry

The Need and the Success

Common sense often makes good law.
—WILLIAM O. DOUGLAS

A law is valuable not because it is law, but because
there is right in it.
—HENRY WARD BEECHER

When everyone expresses their freedom, no one is free
—N. SCOTT MILLS

The American continents have had an interesting interaction with nature since the relatively recent European settling and development of the two western continents. Most importantly, unlike areas in Asia, Europe, and the Middle East, which have been highly populated for millennia, we have watched our continents' nature be assaulted in only a few centuries. This relatively short time frame may have made us more sensitive to this destruction than people on other continents, where it has occurred over tens of thousands of years. Perhaps this is why North America and Europe are given credit for launching the huge environmental movements of the 1960s and 1970s. But when did this heightened environmental awareness really start?

Many countries have the right to clean water and the environment written directly into their constitution, but the United States has taken a much slower approach since its founding. Quasi-environmental issues were a concern for the first European settlers in the Americas, and our laws evolved and are still evolving today. As early as 1626, Plymouth Colony members had enacted laws regarding the cutting and sale of nearby timber. Later, other areas of the United States passed both conservation and preservation measures that specified how much land could be cleared versus how much was required to be left as virgin forest. Much later, laws concerning the preservation of wildlife were passed. Other landmark environmental laws of the 1700s and 1800s included ordinances concerning trash disposal, national parks, tree planting, and other similar natural resource regulations. The 1900s exposed the contrast between conservationists and preservationists and the early policymaking associated with each school of thought. The period after World War II witnessed increased concern with environmental issues, and these concerns brought rapid change in the 1960s. Many newspapers and magazines ran articles related to the environment and two books of the 1960s were particularly prominent: Rachel Carson's *Silent Spring* and Paul Ehrlich's *The Population Bomb*. In addition, either oil spills became more common or, perhaps, their reporting was improved. Either way, the U.S. population took note and became more interested in trying to prevent these environmental catastrophes. In response to the increased publicity and concern about environmental issues, major pollution legislation was passed in the 1960s. Not surprisingly, many new environmental groups formed during this period both to advocate for these laws and to participate in the environmental litigation necessary to enforce them.

The 1970s were even more reactionary because of a variety of direct and indirect factors. Social strife and upheaval, as well as heightened civic-mindedness, were rampant as a result of concurrent racial issues, increased poverty, the radical youth hippie movement, Cold War issues, and even protests at Cape Canaveral about the money supposedly wasted on space exploration (some argued that these funds could more appropriately have been spent on social programs). Images of the issues, including pictures of dramatic environmental disasters, were brought into living rooms every night by Walter Cronkite, as television sets became commonplace in American households. Nothing influences opinions like stark visual images and questions from your children about why these events are happening. I am a product of this generation and I remember these disturbing images vividly.

What was the U.S. response? The bipartisan passage of the most rapidly passed and sweeping set of environmental laws to date, including the Clean Air Act of 1963 and amendments in 1967, the Clean Water Act of 1965, the Endangered Species Preservation Act of 1966, and the National Environmental Protection Act of 1970. During this era, U.S. citizens had coalesced to protest numerous social issues, including the assaults on the environment that they now recognized. Congress and President Nixon, and later President Carter, reacted with regulations. This points to an important distinction to be made between social and political pressure then and reactions today. Back then corporations were not people (my apologies to Mitt Romney!), and really they are not people today.

But first, let us simplify and generalize about politics today. We have two major political parties in the United States, Democrat and Republican, which differ in many ways. One generalization on which probably everyone will agree is that in general, Democrats believe in regulation to solve a problem, while Republicans prefer market-based solutions. Let us go back prior to the 1970s to see evidence of the emergence of ideologies. As mentioned in the introduction, since the formation of the U.S. colonies, numerous environmental regulations have been enacted. Based on the environmental problems that these acts and their amendments sought to address, below is a list of ten, simple, undisputable facts with regard to past industrial practices and the general lack of protection or concern regarding human health and the environment demonstrated by these previous practices:

1. The purpose of the 1948 (and 1972) Federal Water Pollution Control Act was to regulate water pollution. It was replaced in 1965 with the similarly focused Water Quality Act, which was then amended in 1972, 1974, 1977, and 1986 and reauthorized in 1987 as the Clean Water Act. Altogether, this landmark legislation package was passed in response to past improper actions by humans, especially industry, who used our rivers as waste dumps. Examples include our reference to rivers, in chapters 1 and 2, as open sewers and the burning of the Cuyahoga River in Cleveland, Ohio, in 1969.

2. The 1955 Air Pollution Control Act was superseded by the 1963 Clean Air Act with amendments in 1970 and 1990 and by the 1967 Air Quality Act with major amendments in 1977 and 1990. This series of statutes was required to control air pollution on the national level, for not only mobile sources such as automobiles but also especially for stationary, industrial sources. The underlying legislation was passed in response to rampant, uncontrolled, and hazardous air emissions that negatively impacted human health and the environment. The later

versions and amendments to these acts are of particular importance to the cleanup of toxic metal emissions related to coal-fired power plants and the regulation of similarly caused carbon dioxide (CO_2) emissions.

3. The 1965 Solid Waste Disposal Act, amended in 1976, and the 1976 and 1984 amendments to the Resource Conservation and Recovery Act, were passed to address poor disposal practices for solid waste from municipalities and hazardous waste from industry. These statutes mainly affected new landfills and newly generated waste. The sister act, Superfund or CERCLA (the 1980 Comprehensive Environmental Response, Compensation, and Liability Act and 1986 amendments) responded to the need to clean up past municipal and industry hazardous waste sites. These acts affected the past disposal of waste in hundreds, if not thousands, of landfills, most notably Love Canal in New York.

4. In 1969, the National Environmental Protection Act, sometimes referred to as the environmental Magna Carta, was passed to change the mind-set of American industry and government, create greater transparency, and enable more public engagement in environmental decision making, and it served as a national framework to enhance protection of the environment. Its major components were the requirement of an environmental assessment (EA) and environmental impact statement (EIS) for any new projects requiring federal approval, including environmental cleanups.

5. In 1970, the Occupational Safety and Health Act was passed to respond to past abuses in the workplace that submitted workers to unsafe workplace hazards such as excessive chemical exposure and many other hazards. This law governed the safety and health of workers in both private-sector and government workplaces.

6. In 1972, the Federal Insecticide, Fungicide, and Rodenticide Act was passed in response to poor industrial and farming practices regarding the manufacture, sale, and use of agriculture chemicals. This statute required new practices for pesticides and insecticides that would protect consumers, users, and the environment. This act had originally passed in 1947 and was amended as recently as 2007.

7. In 1973, the Endangered Species Act was passed in response to the decimation of sensitive plant and animal species, including the American

bald eagle, due to economic development that destroyed habitats and chemical contamination that affected the health of the ecosystem and harmed various species. A major impetus for the Endangered Species Act was Carson's *Silent Spring.*

8. The 1975 Hazardous Materials Transportation Act was passed as a means of improving safety in shipping, preventing spills during transportation, and addressing illegal chemical waste dumping all in order to protect human health and the environment. Areas of interest included proper tank labeling during shipping, shipping manifests, and banning the shipping of hazardous products in the same vehicle as food items.

9. The 1976 Toxic Substances Control Act (TSCA) and amendments were passed in response to poor practices in the manufacture and import of chemicals. The TSCA essentially regulates the introduction of new and existing chemicals onto the market. Intense discussions have been underway to update this much outdated law to reflect very successful components of the European Union's Registration, Evaluation, Authorisation and Restriction of Chemicals (REACH) program, that require the burden of proof that a chemical is safe back to its industrial manufacturer or supplier. The recent, 2016, passage of the Lauterberg Chemical Safety Act will make progress in this area.

10. The 1990 Oil Pollution Act was passed just after the 1989 *Exxon Valdez* oil spill to better regulate and prevent oil spills. The main improvement provided by this Act was to ensure that a plan was in place to mitigate any future oil spill off the coast of the United States.

As you see from these numerous sweeping bipartisan laws, the social upheaval of the 1960s and 1970s brought about real change. It is also interesting to note how a piecemeal approach to environmental management has finally come to work so well in the United States. It would seem that there was a mastermind behind all of these interconnected and far-reaching Acts, but unfortunately the government had no such mastermind. Rather, there was simply an evolving response by government to demands by citizens and victims for action to protect human health and the environment.

Note the recurring use of the phrase "in response to" in the ten acts listed above. All of our legislative actions on the environment have been in response to poor practices. You could conclude from this that we have a reactionary government and not one of forethought. I refer to this as the "missing traffic light scenario." How does a new traffic light come to be added to a busy intersection? The not-uncommon process is that traffic increases and residents complain to local government; local government conducts a study and tends to delay action because of limited resources; and then a van gets into an accident at the intersection and children are killed. Shortly thereafter a traffic light is finally installed. This is reactionary, not preventive thinking, and it is systemic in our protection of human health and the environment, if not our government altogether.

But after years of abuses, today we have a fairly effective national strategy for regulating practices that affect human health and the environment. And today most of industry supports these actions and Acts. We also have in the Environmental Protection Agency (EPA) a federal entity whose explicit mission is to ensure environmental protection. There are, however, three notable exceptions to our national successes.

First is the issue of potentially harmful chemicals contained in consumer products. While this may seem more of a problem to be solved by the Food and Drug Administration (FDA), when you are talking about a chemical such as bisphenyl A (BPA; see chapter 4) the problem undeniably falls into the environmental arena because of the broader, ecosystem contamination it can cause. Second, is the issue of climate change causing greenhouse gas emissions and the lack of federal leadership, as discussed in chapters 6 and 9. In an attempt to solve climate-change issues we have decided to go back to an age-old argument, state versus federal responsibilities and rights. Many states are stepping up to address climate change (and issues such as BPA exposure) while the federal government is only slowly beginning to act. Today, many states are in fact the centers of innovation on sustainability and are enacting environmental protection measures that are more stringent than existing, or nonexistent, federal laws. Yet states still manage to maintain and even promote economic health. Third is the politicization of the environmental enforcement decisions and budgeting for the EPA. The main problem with our national legal system is enforcement. Unfortunately, too often whether a passed law is actually enforced is left to the discretion of the person sitting in the Oval Office. Especially strong opponents of environmental protection were Ronald Reagan and George W. Bush. While Reagan tried to repeal environmental regulation passed in the 1960s and 1970s in favor of free-market practices, Bush very aggressively tried to undermine environmental safeguards in favor of reducing costs to industry. Here are just a few of Bush's shoddy attempts to undermine science and protection of human health and the environment.

The Bush administration delegitimized science:

- when nonscientists and former oil-industry lobbyists completed unwarranted editing and watering down of climate-changes reports in 2002 and 2003 (National Resources Defense Council 2005; Pearce 2007; Renner 2004; Revkin 2005; Sierra Club 2006),
- when it used false distortions of science to call attention away from the upcoming climate change crisis (*Environmental Science and Technology* 2003; Hogue 2005a; Randerson 2004),
- when it attempted to silence and censor government scientists concerning global warming (Baum 2006; *Chemical and Engineering News* 2005; *Chemical and Engineering News* 2006a, 2006b, 2006c, 2006d; CNN 2008; Hebert 2008; Hogue 2010),
- when it undermined enforcement of well-established environmental governmental acts listed on the previous page (Hogue 2005a; Hogue 2007), and ignored advice from government science panels and advisees (Baum 2006; *Chemical and Engineering News* 2006a, 2006b, 2006c, 2006d; Hess 2006; Petkewich 2006).

All of this undermining of scientific progress, and more, resulted in the resignation of numerous government scientists in protest, not to mention untold harm to human health and the environment. So, as you see, for environmental protection to be effective we need not only good environmental legislation, but also a government that is willing to enforce it.

More in depth, case-by-case examples of misinformation can be found through the Scientific Integrity in Policymaking report by the Union of Concerned Scientists (2004), the website operated by BushGreenwatch (no longer active), the National Resources Defense Council website (http://www.nrdc.org/bushrecord), and the Sierra Club website (2006).

Overall, the United States has excellent environmental legislation in place, but we also need scientifically informed elected politicians who will lead with integrity.

One other issue needs to be noted concerning fines for polluters. In the past, fines have been lower than the damages to humans and the environment. Classic examples come from the mining industry, for which a bond is put into place prior to any mining operation, yet when remediation has to be completed after the mine has closed, or the mining company has declared bankruptcy, the dollar amount of the bond is insufficient to cover the actual cost of remediation. But this is starting to change. While the total damage from BP's *Deepwater Horizon* oil spill is still in question, the total bill to date is approximately $42.5 billion. Likewise, the U.S. government has fined Kerr-McGee $5.15 billion for environmental damage, and subsequent business reorganization fraud, with regard to uranium mines in the Navajo Nation, wood-treating businesses in the Midwest and on the East Coast, and a perchlorate plant near Lake Mead, Nevada. Most agree that the fine was in excess of environmental damages, but it was assessed based on known poor practices and fraud.

And now for the latest in international corporate greed. In September 2015, Volkswagen admitted that 11 million of their diesel vehicles had fake pollution controls programmed into them. The scandal deepened as other automakers were scrutinized, and as two engineers were targeted in the Volkswagen investigation, as Volkswagen's CEO resigned, as the fraudulent software was found to have been used internationally, and as it was suspected that Audi and Porsche also designed faulty diesel SUVs. But international governments, as well as stock markets, resoundingly responded: tens of billions of dollars (and Euros) are coming, Volkswagen's stock price has plummeted, and at least one report (Borenstein 2015) has estimated dozens of emissions-related human deaths resulting from this criminal act. The story will continue to unfold as this book goes to publication, but the take-away message for bad actors in national and international corporations is that they will now, and in the future, pay for their poor environmental decisions.

So far I have addressed only legislation in the United States; as John Donne said "No man is an island" (also made famous to my generation by Grace Slick and Jefferson Airplane). What about global legislation through the UN with regard to pollution, particularly nonlocalized pollution? In the United States, we have run into a problem that is referred to as American exceptionalism. An older, now obsolete entry on Wikipedia defined American exceptionalism as "the proposition that the United States is different from other countries in that it has a specific world mission to spread liberty and democracy. It is not a notion that the United States is quantitatively better than other countries or that it has a superior culture, but rather that it is qualitatively different." (http://en.wikipedia.org/wiki/American_exceptionalism, accessed February 2014). Or to put it another way, how can a foreigner have the insight or authority to tell us, in the United States, what to do?

This attitude falls in line with Neoconservatism and the ideology of the Tea Party. The viewpoint shared by these groups is, what right does the UN or anyone else have to tell us what to do concerning pollution in our sovereign nation or, more radically, on our private property? The problem is air and water pollution—such as acid rain, polychlorinated biphenyls (PCBs) and dichlorodiphenyltrichloroethane (DDT), carbon dioxide, mercury and arsenic from coal-fired power plants, and chlorofluorocarbons (CFCs) from our food-cooling systems—know no national boundaries. Impacts of these emissions, such as climate change, therefore also defy geographic borders.

The United States has been part of two major international efforts with regard to air pollution. The first, the Air Quality Agreement of 1991 was between the United States and Canada and concerned the need to reduce sulfur emissions that caused acid rain, most importantly in the northeastern parts of North America. The second we have already discussed in chapter 6, the Montreal Protocol, which attempts to limit manufacturing, use, and emissions of CFCs that degrade the stratospheric ozone layer. These two instances of international environmental collaboration were an excellent start, but cooperation by the United States with the international community has essentially fallen apart on the environmental front, while the rest of the world moves ahead. As a consequence, Europe, not the United States, is now the global leader in enacting environmental policy. After all, the European Union is an international organization and has continuous experience with working toward consensus and compromise.

Two current, international initiatives come to mind that illustrate this lack of U.S. involvement. The first is the EU REACH program mentioned previously. This program attempts to control the production and sale of chemicals. This is a landmark effort by the European Union to protect human health and the environment and put tight controls on EU industries and chemical imports to the EU. European chemical companies are aggressively working to comply with REACH. If only the United States had a law similar to REACH. All we currently have is the very old Toxic Substances Control Act (TSCA) from 1976; that's 40 years ago! You would think that given the massive increase in chemical production and use we would have updated this by now. And there is call for an update that has support from industry, environmental activists, and both political parties (Hogue 2013) but laws take time, apparently more than 40 years. Finally, after years of debate and negotiation, in June of 2016 Congress passed and President Obama signed the Frank R. Lautenberg Chemical Safety for the 21st Century Act (Erickson 2016), named after the congressman who has fought for years for such a bill. While details of the Act are to be worked out by EPA, U.S. chemical companies will likely operate like those in the European Union.

The second, major international, environmental effort that the United States has abstained from is the SAICM (Strategic Approach to International Chemicals Management). The name says it all. Doesn't it make sense to manage an industry that has become global? Most leaders in the chemical industry think so. Industry always favors an even playing field. Many view SAICM as an easy way to globalize REACH. An international perspective by the United States is absent.

While I have been very blunt about some of our political leaders and the legal process, I view our environmental legislation as a success story. As John Saxe stated in 1869, "Laws, like sausages, cease to inspire respect in proportion as we know how they are made." We have a very complicated, evolving, and effective umbrella of environmental laws in the United States, and these provide environmental protection when they are adequately enforced. We have acted locally in some cases, now we need to join the movement globally. For example, approximately 190 nations, but not the United States, ratified the Kyoto Protocol. All of the U.S. presidential administrations between the Carter and Obama, Democrat and Republican, did little to nothing to engage in global environmental efforts and even damaged international cooperation. These presidents should be held accountable for the upcoming global warming/climate change/climate chaos. While the impacts of the Kyoto Protocol are still limited and it is in the process of being renewed/overhauled, it is shocking that the United States is one of the only Global North nations not to have ratified it (https://unfccc.int/kyoto_protocol/status_of_ratification /items/2613.php). Sixty-three countries, not including the United States, but notably including India, have ratified the 2016 Paris climate agreement. But the U.S. view has been changing under the Obama administration, and this is the subject of chapter 9. Next, chapter 8 discusses how technology and other means will save Earth for future generations of humans and many other species.

EIGHT

The Rapid Advancement of Technology

Our Best Hope

Men are only as good as their technical development allows
them to be.

—GEORGE ORWELL

However far modern science and technics have fallen short of
their inherent possibilities, they have taught mankind at least
one lesson: Nothing is impossible.

—LEWIS MUMFORD

For a successful technology, reality must take precedence over
public relations, for Nature cannot be fooled.

—RICHARD P. FEYNMAN

I'm not sure what solutions we'll find to deal with all our
environmental problems, but I'm sure of this: They will be
provided by industry; they will be products of technology.
Where else can they come from?

—GEORGE M. KELLER

From a reductionist point of view as a physical scientist, I generally believe that our environmental problems can be traced back to two human failings: overpopulation and greed. But some of my colleagues in the social sciences may disagree. Certainly, the world requires a minimum human population to maintain a modern society, but too large a population residing in an area stresses resources, and limited resources then cause greed to rear its ugly head. Capitalism has obvious advantages, but too large a workforce leads to lower wages, and greed leads to excess profits. For years, the global population was predicted to stabilize at approximately 9 billion people because of progressive policies in education and family planning. Unfortunately, recent research has indicated that 9 billion is currently a low, and probably unachievable, goal for now. It is now being predicted that the global population may even reach 11 billion by 2100. Some think that we need to achieve a sustainable global population of 7.7 billion; others believe that technology and agriculture will keep up with the needs of our growing population. But one thing is clear, not everyone on Earth can live at today's standards of Global North residents. But this chapter is mostly about science. For a recent, clear summary of how to reduce our future human population I refer you to Weismann's book (2014), in which he states his belief that global contraception is the answer.

Conservation should come first. Residents of the Global North need to eat lower on the food chain (particularly consume less meat than is currently in the typical American diet) and in the process save freshwater and food resources. Unfortunately, in the near future we will most certainly have to relocate some populations as global climate change devastates some areas of Earth because of impacts such as the rising sea level and massive droughts. Also, we will definitely need to lower our greenhouse gas emissions or Mother Earth will grow tired of us long before we are able to achieve a stable population.

As a teacher, I am very optimistic about the potential of education to inspire behavioral change. In the process of earning their college degrees, students' attitudes become more global. They are encouraged to alter society's course, and in the process, save the world. Today, students are constantly exposed to environmental information, in all academic majors and at public and private institutions alike. Education is our future.

But education does not stop in the classroom. Look at the changes and challenges that society has faced, and mostly overcome, in the past few decades: women's rights, race issues, LGBT rights, gay marriage, HIV/AIDS disease, poverty, and numerous others. Many individuals and groups have helped lead the way for change on these issues, including Hollywood stars, quality news services, the music industry, progressive religious leaders, authors, actors and playwrights, nongovernmental organizations, and a few exemplary political and social leaders. Natural and physical scientists are working to solve the technical side of problems, but we have only recently started to appreciate the role of social scientists in convincing society to accept and adapt these technologies. Important areas in the social sciences include, but are not limited to law, anthropology, behavioral economics, economics, political science, psychology, and sociology. As we have seen throughout this book, modern societies can solve any problem, given the resources and the resolve to implement them. U.S. and international environmental education and research efforts will continue, and countries will increasingly exchange strategies and technological innovations developed for addressing environmental problems.

I see the physical and social sciences as a great hope, but as noted by a few scientists, science and technology alone cannot solve all of our problems (most negatively expressed by Huesemann, 2003). Hence, it is also necessary to stress long-term global population reduction and constant conservation. We need to be patient with science, engineering, and the social science—but only as patient as time will permit—in the development, evolution, and adaptation and acceptance of new technologies. There are many examples of technological innovation and advances. I use figure 8.1 in my classes to discuss game-changing technologies.

Over time, a technology matures and slowly reaches a plateau of development. But it is not uncommon, especially today, for a game-changing advancement in this technology to occur. This elevates the value of the old technology to a new level. And it is increasingly common today for newer game-changing technology to occur, even before the first new technology can take effect, as illustrated later with respect to light bulbs, modern cars, and energy production.

Although they are not a direct link to environmental issues, take computers, for example. I have lived through most of the computer era. I remember using optical punch cards on a very slow main frame; quite a difference from my powerful laptop today! One of the best examples of the speed at which computer technology has evolved is the commonly noted Moore's law governing the speed of advances in processing chips (or in other words, the numbers of transistors that can be placed on an integrated circuit). Computer capabilities have increased rapidly, in fact exponentially, since 1971. The speed of computer chips, which improves

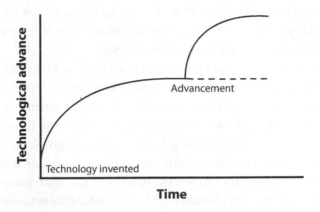

FIGURE 8.1. Technological advancement over time

with each additional transistor, generally doubles every 18 months to 2 years. And recall the many social aspects of computer technology.

Another example of similarly rapid change is innovation in computer data storage devices. Decades ago, we relied on magnetic tape devices; these were replaced by large, low-capacity floppy disks in a sleeve that got small and smaller (and more vulnerable). Next arrived compact disks (CDs), followed by digital video disks (DVDs), and now today we have flash drives the size of a small coin with tens of gigabytes of capacity. But many of our computer successes and environmental advances alike rest on the shoulders of decades-old technologies. First, consider the liquid crystal display (LCD) in your laptop computer; the first observation of the rotation of light by a liquid crystal occurred as far back as the late 1800s. The inventor, Friedrick Reinitzer, was probably told "good luck making money or feeding your family off that invention." But look at all the places it is used today, for example in our new popular big flat-screen televisions. Second, where would all of your electronics be today without the micro-transistor (p-n junctions) invented by Walter Houser Brattain? Third, while microtransistors revolutionized electronics, its impact did not really take off until the early twenty-first century, when inexpensive and large-storage-capacity flash computer memory was invented. My main point in mentioning the above discoveries is that technology takes time to mature and be accepted. We are standing on the shoulders of giants, previous giant scientists. But how much time will Mother Earth give us to tackle our addiction to fossil fuels?

With regard to environmental successes mentioned in this book, realistically our modern water and wastewater treatment systems have evolved over about a century. The relatively simple water and domestic wastewater treatment has certainly reached a plateau, but recent and near-future high-tech solutions will likely solve our problems with pharmaceutical and endocrine-disrupting compounds in our wastewaters. Likewise, technology has allowed the removal of lead from many common products, as well as our current highly successful efforts to remove mercury from industry processes and household products. And relatively inexpensive and simple ways of removing mercury from coal-fired power plants has been developed. Implementation of socially accepted technology will lead our way to an environmentally sustainable future.

Our future technology gains are essentially limitless. For example, automobiles are not commonly thought of as an environmental success story, but they soon will be if we can successfully overcome the negative inertia of the fossil fuel industry. In the 1970s and 1980s, advances in catalysts provided reductions in tailpipe emissions as did on-board computer diagnostics in the mid-1990s. After more than a century, cars are actually becoming less detrimental to the environment, with electric vehicles and hybrid engines. Automotive manufacturers cannot make hybrid cars fast enough, not to mention the new all electric cars. Taking this environmentally beneficial innovation a step further, hydrogen-fuel-cell prototype engines are already being deployed. Cars with these engines produce carbon-free emissions of simple water vapor and are already on the market in limited areas. Our automotive technologies are developing faster then industry can keep up with them. The only impediment to safe hydrogen-fuel distribution is converting, and in some cases replacing, the fossil fuel distribution infrastructure and overcoming the resistance to doing so. But there is one slight problem with current and near-future hydrogen supplies: they will likely come from coal and natural gas. In the long term, the splitting of water into gaseous hydrogen and oxygen will most likely be accomplished with solar energy. I believe in the very near future we will win this battle, assuming that we have progressive national and international government leadership.

Light bulbs are another excellent, relatively recent, example of a game-changing technological advancement that has made positive progress for the environment through enabling greater energy efficiency. Figure 8.2 shows the date of major advances in light bulb technology. This technology follows a similar trend in progress, as it was invented in the early 1800s and has improved with light-speed to the LEDs (light-emitting diodes) on the market today. The energy efficiency of light bulbs, the actual energy converted to visible light and not lost as heat, has increased from 10% to approximately 95%. This improvement will greatly decrease our lighting energy demand, which accounts for a significant portion of our global energy use. Devices such as LED-style light bulbs will have considerable economic and environmental impact around the world. In 2011, the United States used 12% of its total generated electricity for lighting. With complete conversion to LED light bulbs with their 95% efficiency, we could rapidly reduce our energy use and perhaps finally eliminate our reliance on dirty coal. Other countries, where the proportion of electricity used for lighting is greater, could use the electricity saved through energy efficiency to refrigerate food, run advanced water treatment facilities, or develop or grow job-generating industries. And more success is just around the corner. The Energy Independence and Security Act, signed into law by President George W. Bush in 2007, banned by 2012 the manufacture and import of 100-watt incandescent light bulbs, with some exemptions. In 2013, 75-watt light bulbs were banned. On January 1, 2014, 40- and 60-watt light bulbs were supposed to be outlawed, but I still see some of these for sale in department stores. Eventually, as in Europe, there will even be LED replacement light bulbs for fluorescent sources, which will be another win–win advancement, providing more energy efficiency and reducing another source of mercury pollution. Very soon LEDs will not even use rare earth metals imported from China such as phosphorous agents (these turn pure and bright white, red, and blue light from LEDs into warm light). And even more recently, organic warm LEDs are becoming available. Scientists today are making similar advances in many other fields.

These few examples make it clear that technology is one of our best hopes for overcoming modern environmental challenges. However, as mentioned earlier, we also need to allow time for new technologies to develop and be implemented into society. Sometimes this even includes

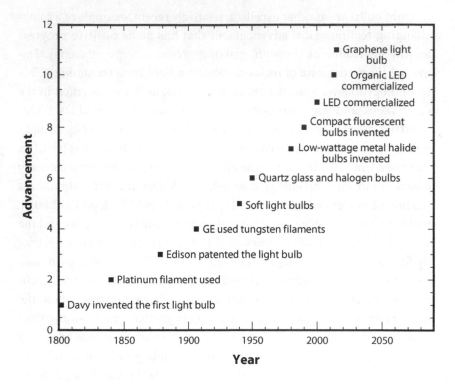

FIGURE 8.2. The light bulb, an example of advances in technology

being willing to consider technologies that may initially be deemed questionable, such as genetically modified organisms (GMOs) and crops and nuclear power. Yes, many people, including many physical and social scientists, think Monsanto's RoundUp® resistant crops are a problematic idea. But safe, well-planned, and well-researched GMO ideas have the chance to greatly reduce our human footprint on Earth, restore ecosystems to their natural state prior to human destruction, and feed the world. Examples include drought- and frost-resistant crops that we already have started developing and fruits with genetically improved shelf lives. There are more of these safe alternatives to be discovered.

Nuclear power has had a questionable run, but many scientists agree that it will be needed to bridge our energy supply until scalable renewable energy facilities are built. And while early attempts with geologic fracking have problems, such as undocumented and possible excessive atmospheric releases and shallow groundwater contamination, fracking will be with us for several years. Newer fracking techniques and better regulations will hopefully allow this technique to supply methane as a bridge fuel to replace coal as we develop more sustainable and nonfossil fuel sources and thereby greatly reduce our carbon footprint. But I want to stress, methane is only a bridge fuel to a fossil-free energy international policy, and fracking technology must be assessed very carefully. When the history of fracking is written, it is possible that the CEOs of a few companies and the politicians that supported them will be guilty of crimes against nature and Mother Earth.

Another optimistic view I have is confidence in the potential of the global liberalism movement, although it is still a slow process. Many nations lead the United States in enacting more progressive policies, but history shows that as time goes on, many societies have become more liberal and tolerant of others, and I would say, more environmentally minded. We no longer burn witches or commit many other egregious acts. And again, Hollywood has certainly contributed to changing political and social views over many decades.

I want to close this chapter with one potentially game-changing technology that could re-create energy production as we know it, the development of nuclear fusion energy. I'm sure you have heard the saying "Nuclear fusion is thirty years away, and always will be." This statement is likely appropriate for the incredibly expensive (multi-billion dollar) price tag of the research fusion reactors in Europe and China that may or may not eventually lead to a sustainable fusion reaction and power generation. But recent ingenuity and technological developments may put an end to the fusion joke above. When Lockheed announced its breakthroughs in small reactor design and plasma containment in October 2015, the energy world took notice. This development has spawned numerous news reports, but the promise was best summarized in a recent issue of *Time* (Grossman 2015). Relatively low-cost, that do not produce radiation or radioactive by-products, and with no risk of containment loss, redefines the meaning of a win–win scenario. I encourage you to read this descriptive article that could make the fossil fuel industry the next species to become extinct.

Since the 1970s, we have made incredible advances in environmental health, and while at every turn industry screamed "economic doom," today, industrial entities have developed the technologies required to implement the regulations. I should also note the combined development of technologies through governmental regulations and free-market enterprise—yes, good old capitalism. Countries that have implemented these technologies generally have the cleanest environments and the greatest economies. As noted in this book, we have cleaned up our rivers and drinking water, have eliminated human exposure to many toxic and heavy metals, are starting to address animal (including human) endocrine disruptors, have very valid approaches to risk assessment in place, have addressed three major atmospheric pollution problems (smog, acid rain, ozone depletion), and have established an excellent set of environmental laws. Unfortunately, what may be humanity's greatest challenge is climate change, which is the subject of chapter 9.

NINE

Humans' Greatest Challenge

Climate Change

Two thousand scientists, in a hundred countries, engaged
in the most elaborate, well organized scientific collaboration
in the history of humankind, have produced long-since a
consensus that we will face a string of terrible catastrophes
unless we act to prepare ourselves and deal with the
underlying causes of global warming.

—AL GORE

All across the world, in every kind of environment and region
known to man, increasingly dangerous weather patterns and
devastating storms are abruptly putting an end to the
long-running debate over whether or not climate change is
real. Not only is it real, it's here, and its effects are giving rise
to a frighteningly new global phenomenon: the man-made
natural disaster.

—BARACK OBAMA

Though more work always remains, the physical sciences have
accomplished their core task when it comes to climate change.
We know what we need to know about the causes and
consequences of our actions. What we don't know is how to
stop ourselves, which is why this book—and the social
sciences—are so important from here on out.

—BILL McKIBBEN, *THE END OF NATURE*

These three quotes refer to the greatest challenge humans face, and
likely ever will—global warming, climate change, climate chaos, or whatever you want to call it. While now 99% of researchers in the climate
change community have come to consensus on the fact that this problem

is real, is happening, and is caused by humans, incredible controversy surrounds the last issue—controversy driven by the fossil fuel industry, who stand to lose profits. But I want to especially note the third quote. The science and causes of climate change are essentially conclusive. Now we must let the social scientists do their work in constructing a viable plan for climate change mitigation and for the short-term adaptation to upcoming climate changes. Hence, social scientists are a substantial part of the Intergovernmental Panel on Climate Control. But I am a physical scientist, and in this book I will report only on the science behind anthropogenic-caused climate change.

This chapter is in part a continuation of chapter 6, the final environmental danger to Earth's atmosphere. While not a success story now, hopefully for our sake and the sake of many Earth inhabitants, not just humans, in the next few decades we will turn away from this dangerous corner. First, let us examine some recent reports from the scientific literature and popular news.

- Our oceans are turning acidic faster than they have for the past 300 million years (Parry 2012).
- Arctic ice has shrunk to an all-time historic low; it has decreased in size by 50% since 1980 (Borenstein 2012).
- Even if we successfully acted today, the Western Antarctic Ice Sheet containing the Thwaites Glacier will melt over the next two hundred years and by itself will account for a 2-foot rise in sea levels.
- It is estimated that "fossil-fueled energy production . . . causes $120 billion worth of health and non-climate-related damages in the US each year" (Johnson 2009c). This would result in a 25% higher energy bill if these costs were factored into the retail cost of electricity (Johnson 2009b). Yet, atmospheric concentrations of carbon dioxide (CO_2), the major contributor to global warming, continue to increase. Global CO_2 emissions set a record high in 2011 (and will likely each coming year as well) of 31.6 billion metric tons (Hogue 2012).
- The summer, and likely the year, of 2015 was the latest in a series of warmest years on record, the sixteenth driest year on record, and the second most extreme weather year on record for the United States.

- In 2013, the average concentration of CO_2 in the atmosphere reached the much-feared mark of 400 parts per million (Gillis 2013), but this is likely only one of many much-feared milestones to come.
- Not surprisingly, intelligence agencies have been advised to plan for global-warming disasters (Johnson 2012).

These are the results of scientific research, not political opinion or left-wing rhetoric. Figure 9.1 gives you insight into how climate change will affect your region.

If you want to know what the world will look like if all the ice melted, go to the Sea Level Rise Viewer tool on the National Oceanic and Atmospheric Administration (NOAA) website (https://coast.noaa.gov/digital coast/tools/slr). Warning, these are depressing images.

This chapter will start off with a brief introduction to the science of climate change—scientific facts, not opinion. I will not go into great detail here since, as noted above, you likely already have formed an opinion on the human influence on climate change. Everyone, deniers and concerned climate advocates alike, can learn from reviewing the key data that provide the evidence. Numerous sources of reliable information exist on this topic, ranging from textbooks to popular novels. The short background that follows is necessary, however, to set the stage for two data sets that are argument-deal breakers for the adamant deniers of climate change.

Our atmosphere has been relatively stable, with respect to chemical composition and temperature, for approximately 150 million years. Yes, the concentrations of minor components such as CO_2 and methane have shifted slightly, and these changes have subsequently caused small changes in atmospheric temperatures. This general atmospheric stability, however, has allowed life to evolve and flourish on Earth. The composition of dry air is 78.09% nitrogen (N_2), 20.95% oxygen (O_2), 0.93% argon (Ar), and 0.039% CO_2, with smaller amounts of other gases. Water vapor varies considerably, but an average value is approximately 1%. Of these listed natural gases, only CO_2 and water vapor contribute to global warming.

Most of the radiation (light) coming in from our Sun is in the form of ultraviolet, visible, and near-infrared (near-IR) wavelengths. As previously discussed, most of the incoming ultraviolet (UV) radiation is removed by the ozone layer. Some of the incoming visible light is reflected back

	Region	Description
	Northeast	Communities are affected by heat waves, more extreme precipitation events, and coastal flooding due to sea-level rise and storm surge.
	Southeast and Caribbean	Decreased water availability, exacerbated by population growth and land-use change, causes increased competition for water. There are increased risks associated with extreme events such as hurricanes.
	Midwest	Longer growing seasons and rising carbon dioxide levels increase yields of some crops, although these benefits have already been offset in some instances by occurrence of extreme events such as heat waves, droughts, and floods.
	Great Plains	Rising temperatures lead to increased demand for water and energy and impacts on agricultural practices.
	Southwest	Drought and increased warming foster wildfires and increased competition for scarce water resources for people and ecosystems.
	Northwest	Changes in the timing of streamflow related to earlier snowmelt reduce the supply of water in summer, causing far-reaching ecological and socioeconomic consequences.
	Alaska	Rapidly receding summer sea ice, shrinking glaciers, and thawing permafrost cause damage to infrastructure and major changes to ecosystems. Impacts to Alaska Native communities increase.
	Hawai'i and Pacific Islands	Increasingly constrained freshwater supplies, coupled with increased temperatures, stress both people and ecosystems and decrease food and water security.
	Coasts	Coastal lifelines, such as water-supply infrastructure and evacuation routes, are increasingly vulnerable to higher sea levels and storm surges, inland flooding, and other climate-related changes.
	Oceans	The oceans are currently absorbing about a quarter of human-caused carbon dioxide emissions to the atmosphere and over 90% of the heat associated with global warming, leading to ocean acidification and the alteration of marine ecosystems.

FIGURE 9.1. The effects of climate change around the United States. *Source*: U.S. Global Change Research Program (http://www.globalchange.gov/sites/globalchange/files/NCA3-climate-trends-regional-impacts-brochure.pdf).

into space, but the majority of it reaches the Earth's surface. Meanwhile, some of the near-IR radiation is absorbed by water and CO_2 in the atmosphere. The UV, visible, and near-IR wavelengths that do hit the surface are absorbed and converted into heat, known as near-IR and far-IR wavelengths, which are then reflected back up toward space. At this point, one of two things can happen. If the atmosphere is low in CO_2, water vapor, and other greenhouse gases (such as methane), very little of the heat wavelengths will be absorbed, the atmosphere will warm only slightly, and the majority of the reflected and outgoing radiation will go back into space. However, when the concentrations of CO_2, water vapor, and other greenhouse gases are relatively high, the gaseous molecules will absorb more of the IR radiation and their bonds will vibrate faster. This higher concentration of greenhouse gases and resulting greater absorption of outgoing radiation will heat the atmosphere. This has happened to some extent in the past, is happening today, and is expected to increase in the near future. A certain concentration of natural greenhouse gases, creating a desirable greenhouse effect, is essential to make our planet habitable. It keeps our climate at a temperature that has allowed life to evolve and thrive. The much higher and continually increasing concentration of greenhouse gases contributed mainly by human industrial processes, however, trap additional heat in the atmosphere that is causing our planet to warm at a rate never before experienced.

Scientists have known the connection between increasing concentrations of greenhouse gases and a warming atmosphere for some time. As far back as 1896, Svante Arrhenius (Arrhenius 1896) proposed that the release of CO_2 from fossil fuel combustion would cause significant global warming. More recently, extensive data gleaned from ice cores has allowed us to examine the climate back 650,000 years. We will discuss these data in detail later, but the connection between elevated levels of atmospheric CO_2 and methane and increasing global temperatures, when astronomical factors related to the relative position of the Earth to the Sun have been factored out, has been scientifically proven. But is all global warming bad? It depends on the degree.

As previously mentioned, scientists know that habitable temperatures on Earth result from the presence of limited amounts of global warming gases in our atmosphere. For example, relatively low concentrations

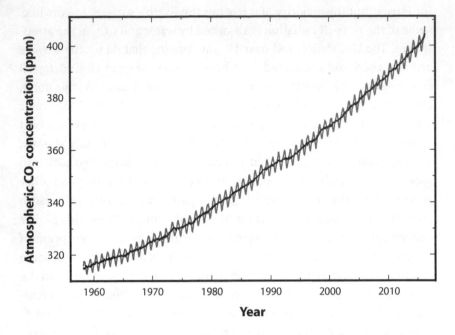

FIGURE 9.2. Atmospheric CO_2 at Mauna Loa Observatory. The Keeling data set shows the annual CO_2 cycle and increase in recent decades. *Source*: Earth System Research Laboratory (http://www.esrl.noaa.gov/gmd/ccgg/trends/#mlo_full); see also Keeling et al. 1976 and Thoning et al. 1989.

of greenhouse gasses in the atmosphere coincided with ice ages. So the real issue of greenhouse gas concentrations is one of the magnitude and rate of atmospheric composition change.

In the late 1950s, the U.S. government started monitoring CO_2 levels in Hawai'i in order to capture average annual concentrations in the Northern Hemisphere and any changes that might be occurring. A recent plot, known as the Keeling data set, is shown in figure 9.2.

As you can see clearly, CO_2 concentrations have been on a steep rise since monitoring began in 1958. But geologically speaking, this is a very short time frame in which to study a planetary event. So let us look further back in time. Historical data have been obtained by drilling deep into glacier ice; as you advance deeper into the ice, you are going back into time. If sampling is done correctly, the gases stored in the ice can be trapped, enabling quantification of historical levels of greenhouse gases such as CO_2 and methane. These data can be combined with data on oxygen isotopes that allow estimation of temperature changes at various

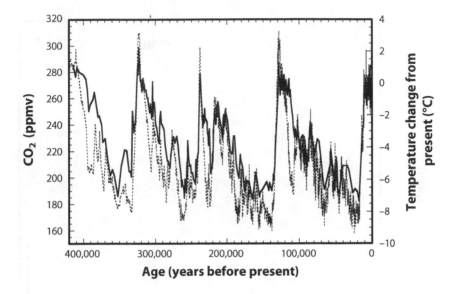

FIGURE 9.3. Prehistoric measured CO_2 and estimated atmospheric temperatures. *Source:* National Oceanic and Atmospheric Administration (http://www.ncdc.noaa.gov/paleo/icecore /antarctica/vostok/vostok_data.html).

points in history. Such data sets have been collected several times and at several locations, and they all show the same data trend: relatively high CO_2 and methane concentrations coincide with relatively high atmospheric temperatures, while relative low CO_2 and methane concentrations coincide with relatively low atmospheric temperatures. One such data set, with data for the past 400,000 years, is shown in figure 9.3. The figure shows the corresponding CO_2 concentrations to relative increases or decreases in the atmospheric temperature.

The peaks in this graph coincide with warm Earth temperatures while the lows coincide with ice ages. Based on this graph, ice cores have enabled us to obtain a very long historic record of the relationship between Earth's CO_2 concentration and the temperature of the atmosphere. When you put the Keeling and ice core data sets together you get a very scary picture of what we are doing to the atmosphere, as shown in figure 9.4.

First, note the bottom axis figure 9.4. This plot goes back 400,000 years. Large peaks in CO_2 concentration that are visible at several points indicate warm periods in Earth's history, but these are dwarfed by the scale

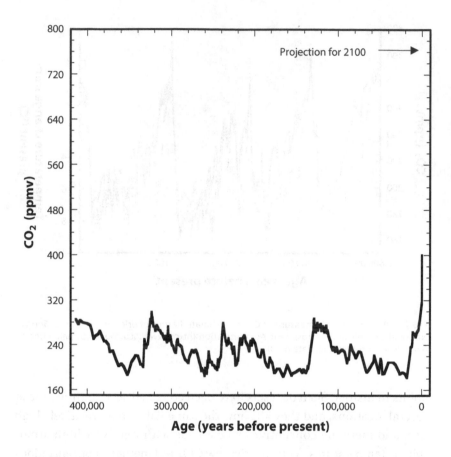

FIGURE 9.4. Combined data from figures 9.2 and 9.3 show a projected level (~750 ppm CO_2) for the year 2100 if no action is taken to curb our CO_2 emissions

on the right-hand side of the plot, which is necessary to plot both today's data and anticipated future levels of CO_2 in the atmosphere. Our current atmospheric CO_2 concentration is higher than it has been in the past 800,000 years based on other studies, at 400 ppm. Of more importance to the heightened levels of CO_2 today is the rapid increase in change (rate), which based on the ice core data, is unprecedented. Atmospheric CO_2 concentrations are changing at a rate never before observed in geologic history. Even more disturbing is the predicted future concentration in 100 to 200 years of 750 to 1,000 parts per million by volume (a corresponding 6°C increase in temperature), shown at the top of the plot if no action is taken to reduce human emissions of greenhouse gases. At this level, our planet will cross a tipping point in which methane, a potent greenhouse gas, will be released from the oceans and from the long-frozen seabed north of Siberia (permafrost). At the same time, white snow-covered polar regions will melt, and the darker Earth surface will absorb more radiation and release more heat to be trapped by the atmosphere, as detailed in *The Hockey Stick and the Climate Wars* by Michael E. Mann (2013). If this plot does not scare you, or if you still deny the existence of human-caused climate change, then there is nothing more I can say to convince you otherwise. Human-caused global warming *is* real. It is here today, it is already causing climate change, and if unabated, it will soon cause even more climate chaos. We can choose to regulate fossil fuel emissions or Mother Earth will choose to regulate her inhabitants.

There are three main points that summarize the crux of the climate change problem. First, global warming, climate change, climate chaos, or whatever term you use to refer to it is real. It is here, today, not a prediction of the future for our children or grandchildren. The impacts of climate change will, however, be even worse for future generations. It is relatively simple to connect the dots between increased anthropogenic emissions of greenhouse gases and climate change as discussed in the above list, especially since no known natural cause can explain our warming planet and observed climate changes. So why are some people still debating whether this problem exists and if it does, whether it is caused by humans? Some climate change advocates, who are very smart, rational, and pragmatic, think that they can still use logic and conversation to overcome greed. An increasing number of scientists and educated policymakers, on both sides of the political aisle, are finished with the debate, accept climate change as real, and no longer tolerate input from adamant climate-change deniers.

The second point is a message to unproductive and highly vocal deniers of science, the leaders of the climate-change–denier movement, the fossil fuel industry and their hired politicians, or as they are most accurately referred to "members of the flat Earth society." This name is applicable to this group because no matter what evidence they are provided with, they will not admit that human actions are responsible for climate change just as opponents to the understanding that the world was round dogmatically held on to their beliefs that the Earth was flat. The leaders of the flat Earth society are not stupid. They completely understand the connection between fossil fuels, CO_2 and methane emissions, and climate change (see the undisputed data plots presented in the figures). The problem is that these individuals have a vested interest in the massive, multi-billion-dollar, global industrial complex that will have to fundamentally change in order to eradicate the causes of human-influenced climate change. Their policy strategy was largely adopted from the tobacco and leaded gasoline industry as was laid out in chapter 3 and clearly and thoroughly discussed in *Deceit and Denial* by Markowitz and Rosner (2013) and *Merchants of Doubt* by Oreskes and Conway (2010). Overall, the fossil fuel industry's opinion of climate change has moved through the following stages of denial (taken from Carey 2007). First, "There's no evidence that the climate is changing." Then later, "Well, maybe it is, but humans aren't to blame." According to Carey, some also argue that, "The warmer the better, and we can easily adapt." Others have even stated that God would not allow this to happen to us. Another more recent rationale is that plants will grow better and faster, in certain limited areas of the Earth, with more atmospheric CO_2 and that the planet will adjust. And the latest argument for inaction is that the U.S. job market and global economy cannot handle the cost of changing our energy system. We will review the cost, or actual benefit, of change later in this chapter. Of course, all of these are lies that have been proven wrong in the scientific, social, and economic peer-reviewed literature. Scientific data clearly demonstrate the link between climate change and human actions. Most current life forms will not be able to handle the increased temperature caused by increasing CO_2 levels. Many economists have shown that the long-term economic costs of not acting are in fact greater than those of inaction, and we need to begin to mitigate climate change now. In fact, our economy, job market, and global health will actually be better and stronger with action. Simple greed drives the members of the flat Earth society. Remember Jamie Kittman's mantra from chapter 3 and the tetraethyl lead argument, "Profit at any cost." The relevant cost in terms of climate change is the cost to life on Earth as we currently know it.

The third point is a continuation of greed, but this may cause the terminal downfall for many climate-change deniers in coal-rich states. Similarly, in general, these same individuals have fought teaching evolution in public schools, but they have lost this argument to science. The Global Warming argument is very similar—science against political and religious ideologies. How can these deniers possibly change their platform now after arguing for decades against the science of climate change? First, there is still a lot of money to be made by the news media on advertising by the fossil fuel industry and by politicians in the form of speaking engagements and political campaign donations. But as the decades go by and climate chaos and associated effects become more obvious, how will our political parties explain their position? Their continued denial of climate change may be their doom. An interesting situation occurred in the summer of 2015 when Pope Francis entered the argument, from a humanity and social perspective. When the Pope called for a transformation of our lifestyles to address climate change, he was quickly criticized by climate-change deniers, who claimed that this was a political argument and not a religious one, and that the Pope had no right express his opinions since he had no science training. (Of course, this is *not* a political argument, but one contrasting greed and science.) The Pope had the perfect comeback when he reminded them that he had a bachelor's degree in chemistry. Pope 1, climate change deniers 0.

Average skeptical U.S. citizens are the true victims of adamant climate-change deniers, since they are lied to by their leaders. Note that I have limited my discussion to U.S. citizens. An amazing percentage of the world, minus oil executive and political leaders in oil-producing countries, of course, openly accept human-caused global warming as real and proven. In the United States for some reason, this has largely turned into a debate between the political left and the political right. Remember that despite the polarization today, it was in fact a Republican, President Richard Nixon, who started the political environmental movement by signing key environmental legislation in the 1970s. In addition, recently 300 major companies agreed to support climate change efforts. Sociologists have shown over and over that when faced with difficult decisions, people tend to turn to their comfort zone for guidance. In very general terms, liberals tend to turn to MSNBC, middle of the roaders watch CNN, and conservatives often rely on Fox News and Rush Limbaugh. People will tend to believe what they want to believe, no matter what the evidence to the contrary. Unfortunately, the average skeptic when it comes to climate change is the victim of propaganda and lies generated by the greed of fossil fuel companies and the politicians they sponsor. Fortunately, more and more people are starting to understand climate science via television news and recognize that data set after data set, competent scientist after competent scientist, and study after study clearly show that human-caused climate change is real. Or even worse, they have already been victims of violent climate change.

Given the scale and importance of climate change—the end of life on Earth as we know it—the United Nations (UN), at the request of member governments, created a scientific, intergovernmental body explicitly focused on this problem, the Intergovernmental Panel of Climate Control (IPCC). The IPCC is tasked with providing comprehensive scientific assessments of current data on the risks posed by climate change and on how human activity is contributing to the problem, aside from natural climate fluctuations. The IPCC recruits thousands of physical and social scientists, and experts from numerous disciplines and sectors complete these reports, all without compensation from the IPCC. This is one of the largest and most technically qualified committees in the world, but since we are talking about the future of life on Earth, this type of collaborative effort fits the nature of the problem—if only our domestic institutions could come together in the same way. The 2014 summary of these reports is subject to line-by-line approval of all UN governments, which usually includes at least 120 countries (www.ipcc.ch). The Fifth Assessment Report and final report was issued in early 2014 and each report since the first in 1990 has issued progressively worse predictions. The final synthesis report was released in November 2014 (http://www.ipcc.ch/pdf /assessment-report/ar5/syr/SYR_AR5_FINAL_full_wcover.pdf) and the key conclusions include (clearly summarized by Fischetti 2014).

The causes of climate change are concluded to be as follows, quoted from Fischetti (2014):

- "CO_2 emissions are by far the largest cause of global warming and ocean acidification, and they are rising.
- Methane emissions are the second largest cause of warming, and they are rising.
- Since 1950, human activities have led to virtually all temperature rise.
- Natural forces have caused virtually none of the temperature rise.

- The largest human sources of CO_2 emissions are burning fossil fuels, making cement, and burning off gas ('flaring') from oil and gas production."

The impacts are:

- "Sea level is rising, and at an increasing pace.
- Glaciers are melting, ice sheets are thinning, and Arctic sea ice is disappearing.
- Permafrost is thawing.
- In North America, snow pack is decreasing.
- The number of cold days and nights are decreasing.
- The number of hot days and nights are increasing.
- Heat waves will occur more often and last longer.
- Heavy rainstorms and snowstorms will become more intense and frequent.
- Overall, precipitation will rise in high latitudes and the equatorial Pacific. In mid-latitudes, dry areas will get drier, wet areas will get wetter.
- Species are vanishing at an alarming and ever-increasing rate.
- Most plants, small mammals and ocean organisms cannot adapt fast enough to keep up with changes.
- Global temperature rise greater than 2 degrees Celsius will compromise food supplies globally.
- Human health problems will get worse.
- Risks to poorer people are greater than for others, in all countries."

What we must do:

- "To avoid severe damage to natural and human systems, the world should keep global warming to less than 2 degrees C above pre-industrial levels.
- Without more mitigation than is being done today, the temperature is more likely than not to rise by 4 degrees C by 2100.
- Significant reductions in greenhouse gas emissions by 2050 can significantly reduce warming by 2100.

- Keeping greenhouse gases in the atmosphere below the equivalent of 450 parts per million of CO_2 can keep warming below 2 degrees C.
- Levels are likely to stay below 450 ppm if human emissions are reduced 40 to 70 percent by 2050 compared with 2010 levels.
- Allowing levels to reach 530 ppm by 2100 gives the planet slightly better than 50-50 odds of staying below 2C; that would require reducing emissions 25 to 55 percent by 2050 versus 2010.
- To hit a target of 430 to 530 ppm by 2100, the world must invest several hundred billion dollars a year in low-carbon electricity sources and energy efficiency.
- It is highly unlikely the world will stay below 450 ppm without widespread use of carbon capture and storage technologies."

These statements are clear and confident. This work was endorsed by another internationally recognized society, the Norwegian Nobel Committee, when they awarded the Nobel Peace Prize to the IPCC in 2007. You may also remember another recipient of this award, Vice President Al Gore for his well-known work on climate change. Members of the flat Earth society are in the minority in the United States and especially in the international community.

But what about the very small handful of supposed scientists who dispute these climate data, including the UN's IPCC conclusions and recommendations? Leaders of the flat Earth society and their media puppets like to compare the scientists who support their views to historic scientists who made breakthroughs against popularly held beliefs, such as the once radical, but now known to be correct, ideas of Galileo and Copernicus. But one huge flaw with these analogies is that radical scientists such as Galileo and Copernicus were scientists using scientific methods and scientific facts to dispute traditional, and largely superstitious, beliefs. And several recent investigations show that many scientists who are denying climate change receive funding from the fossil fuel industry or their foundations. I would equate a scientist who does not believe in human-caused global warning today to a chemist who does not believe in atoms, a biologist who does not believe in evolution, a physicist who does not believe in quantum theory, or a geologist who does not believe in plate tectonics. Outside of a profession in science, it would be like a religious leader who does not believe in their god.

In contrast, competent scientists are in considerable agreement. As recently pointed out by competent news organizations (and John Oliver has an excellent piece at http://www.theverge.com/2014/5/12/5709420/john-oliver-skewers-cable-news-and-climate-skepticism), typical debates that pit a climate-change denier against a climate-change believer are skewed. The debate is not a one-on-one debate, but if statistics were accurately portrayed there would be only three climate-change deniers at the debate table as compared with 97 technically trained climate-change scientists. The point is that 97% of the scientists agree that humans are significantly contributing to global warming and that global warming is occurring now. A more recent scientific survey of climate-change researchers, involving 24,000 peer-reviewed scientific papers, put the agreement at 99.99%. A few scientists do urge caution about the degree of certainty that can be attached to these statements, but this number grows smaller every day.

I will continue my factual rant with the following analogy of the flat Earth society. Say you have a dire health issue. Where would you go? Likely somewhere like the Mayo Clinic, where you would be diagnosed by the most educated, most experienced, and most respected physicians in the world, similar to the researchers at the IPCC. They would give you their analysis and advice and then it would be up to you to decide how to proceed. If you were a climate-change denier, you would promptly ignore their advice and treat your symptoms based on your personal beliefs rather than on the insights provided by medical science.

So where have U.S. policymakers stood on this issue? The minority of well-informed individuals have simply been outnumbered by the well-funded political voices of climate-change deniers. Presidents Richard Nixon and Jimmy Carter were our first, and until recently with the rise of President Obama's environmental policies, our only environmental presidents. Many scientists feel that we missed our opportunity to act in time to avoid or greatly minimize global warming during the Reagan administration. Politicians, Democrat and Republican, at that time and up until today, however, chose to use the "need for more study" card to delay action, sometimes in exchange for campaigns funded by the fossil fuel industry. At times, government scientists trying to speak out about climate change have essentially been censored. There are other instances when technical documents on global warming that were originally written by scientists have been edited by political bureaucrats.

Finally, attempts have been made to discredit and even slander academic scientists, such as Michael Mann and Naomi Oreskes, who sought to offer peer-reviewed research on climate change. Even as of November 2014, the House passed a bill forbidding scientists from advising the Environmental Protection Agency (EPA) on their own research (Abrams 2014). Europe has now taken the lead on international environmental policy, while in the United States we are still arguing and playing catch-up. To summarize, as stated in the movie *American Beauty*, "Never underestimate the power of denial." And I would add, "the power of greed."

Here is another sobering thought. It took years in the 1960s to 1970s for acid rain to be widely recognized as a problem. It was not until 1991, with the amendments to the Clean Air Act, that legislative action occurred, and we are still in the process of installing the requisite pollution abatement equipment on old coal-fired plants in the United States. China is still far behind with remediation but is reportedly completing one new coal-fired plant each week, but the Chinese are actively researching and have the world's largest CO_2 sequestration plant. Similarly, the stratospheric ozone hole was well documented in the late 1970s to early 1980s, but significant global action did not occur until the passage of the Montreal Protocol in the early 1990s. It was only after this international agreement was secured that the phaseout of chlorofluorocarbon (CFC) production and use began. As previously discussed, the ozone hole is not expected to recover until 2050 to 2065. The point is, atmospheric remediation takes a very long time because of the lengthy, multistep, and often delayed process; first, the problem must be recognized and researched, then international consensus and cooperation must be fomented, and finally the atmosphere must be given time to recover. Global warming due to human action was anticipated in the mid- to late 1800s. Over two hundred years later, we still have not acted to significantly stabilize, much less reduce, fossil fuel emissions. With dedicated action today, it will still take one to two centuries for our atmosphere to recover to preindustrial CO_2 levels. And a few scientists feel that it is already too late.

But what sectors of industry are the most responsible for emissions of greenhouse gases? This is where we must immediately act. EPA has just focused on who's to blame in the United States for CO_2 emissions and the plot that follows clearly summarizes where we need to take action: power production, most notably coal-fired power plants (EPA 2015).

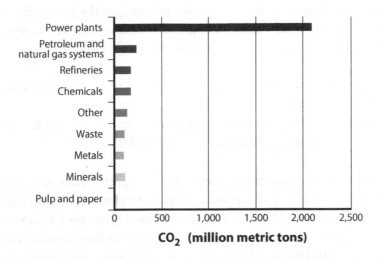

FIGURE 9.5. U.S. greenhouse gas emissions by sector in million metric tons of CO_2. *Source:* EPA 2014.

Coal- and fossil-fired power production dwarfs all other industrial sectors. While in the United States, Europe, and India coal mining and use has stabilized and should hold steady for the coming decades, when will China also stabilize? There was early evidence in 2013 that China was making first attempts, such as investing in renewable energy and large CO_2-sequestration projects. And corruption between the leading Chinese politicians and the oil industry has started being investigated (Forsythe 2014). Containing China's emission has been a major concern from U.S. industry leaders, who state that it would be futile for the United States to take unilateral action if China is to take no action. Enter the United States–China agreement of November 2014 that promises to (https://www.whitehouse.gov/the-press-office/2014/11/11/fact-sheet-us -china-joint-announcement-climate-change-and-clean-energy-c; and Associated Press 2014):

- "reduce U.S. greenhouse emissions by 26 percent by the year 2025 (based on 2005 emissions),
- double the rate that the U.S. is reducing its emission per year,
- submit the new 2025 emission reduction target to the U.N. Framework Convention on Climate Change . . . in 2015 in Paris."

In turn, China will (https://www.whitehouse.gov/the-press-office/2014 /11/11/fact-sheet-us-china-joint-announcement-climate-change-and -clean-energy-c; and Associated Press 2014):

- "set a target for carbon dioxide emissions to peak by around 2030, . . . earlier if possible
- expand the share of China's energy consumption derived from zero-emission source to about 20 percent by 2030."

The fossil-fuel industry's response was as expected, unproductive, and self-serving. But on a surprisingly positive note, the U.S.–China agreement has enticed India to reflect on and possibly reverse its planned increase in coal production. A pivotal moment may have come in August 2015 when the Obama administration ordered power plant owners to cut CO_2 emissions by 32% by 2030. This will certainly play out in the courts, but the U.S. Supreme Court has already ruled that the EPA has the right to control the emissions of CO_2.

But when, or if, we can adequately reduce atmospheric CO_2 levels to those that scientists estimate are acceptable (approximately 350 ppm), how long will it take for the climate to stabilize? Even if we magically fixed the problem now, it is estimated to take at least one hundred years for the atmospheric CO_2 levels to return to acceptable concentrations. Does this mean that it is too late to act? No, it means we need to double down on our efforts. One study showed that we need to start decarbonization (the removal of CO_2 emissions to the atmosphere) efforts now, at a rate of 6.2% per year to minimize our impact on global warming to only a 2°C increase by 2100 (Harvey 2014).

So, given our current inaction, what is expected to happen to life on Earth as global warming continues? Since climate change is here now and will only get worse, the effects are already visible and widespread across society and ecosystems. World Bank economists estimate that by 2050 damage to global gross domestic product (GDP) will create losses of 5 to 20%. On the other hand, if mitigation action is taken sooner, the annual GDP is anticipated to decline by only 1%. More recently, the IPCC has predicted cost estimates of inaction and a 2°C temperature increase to be between 0.2% and 2% of GDP (Bump 2014). In contrast, the cost of acting today would only cost 0.06% (Ritter 2014). However if we do not act and the atmosphere temperature increases 4°C or more, the economic cost will range from 5 to 20% of the GDP, not to mention the human suffering and species extinction (Bump 2014). CO_2 emissions are warming the oceans and will likely alter ocean circulation patterns (the ocean currents made famous in the movie *Finding Nemo*). This will greatly affect climate patterns, ocean life, and ocean food-chain productivity. It is estimated that by 2050 food prices will double, and in some regions there will be considerable food scarcity. Sea levels could rise by as much as a meter (3 feet) in certain areas by the end of this century, and more recent estimates put this rise at tens of feet. Coastal areas will also be affected by swelling ocean surges, tropical storms, and hurricanes. Destructive storms like Hurricane Sandy and Typhoon Haiyan and the polar vortexes of the winter of 2013–2014 will most certainly increase in frequency. We will have no choice but to force mass migrations from highly affected areas of the planet. Regional wars will be more likely as water and food resources become scarcer, combined with massive ecosystem disruption and species extinction. Less than 1% of our global GDP to mitigate global warming does not seem like a high price to pay if we act now to avoid all of this chaos and more.

The good news is that everyone is finally taking notice. I know of no popular magazine that has not had at least one article on global warming/climate change. Along with evolution, climate science is finally taught in secondary education, and even earlier in some school systems. Early in 2014 the IPCC released its Fifth Assessment Report and did a lot of public relations announcements with this release; the IPCC synthesis report is going farther. Social scientists are becoming heavily involved in our climate-change mitigation efforts. In addition, President Obama became active with respect to global warming and has taken a U.S. lead on action to minimize global warming with his Climate Action Plan. President Obama pledged his final two years in office to combating global warming. And finally, the verdict is in: the U.S. Supreme Court has ruled that the EPA has the right under the Clean Air Act to regulate CO_2 emissions, as well as other emissions from coal-fired power plants. An excellent article by former U.S. vice president Al Gore, entitled "The Turning Point: New Hope for the Climate," is definitely worth a read.

On June 2, 2014, Obama's EPA announced upcoming rules to limit and control CO_2 emissions from power plants. These rules consist of cutting emissions by as much as 30% from 2005 levels by 2030 at a cost of only $5.5 billion, yielding a health benefit of between $26 billion and $45 billion. The political wing of the fossil fuel industry was fast to respond, calling the plan "nuts" and a "national energy tax" (http://www.speaker.gov /press-release/boehner-responds-president-obama-s-national-energy -tax). Another referred to the plan as a "dagger in the heart of the American middle class" (Cama 2014). No surprises there. But now that the United States has committed to action, and with the U.S.–China agreement, there is hope that China, and even India, will make considerable efforts to curb CO_2 emissions.

How long will it take for global consensus on a plan and then to put the agreements into action? And how long will it take for the atmosphere to positively respond to our remediation efforts? The Supreme Court rulings can set a series of actions by the EPA into place. In my opinion, in the short term the EPA could mandate conversion from coal- to methane-fired power plants. The rationale is simple: energy generated from methane has half the climate-change potential (effective CO_2 emissions) as coal. In the long term, say 5 to 10 years, the EPA could require all fossil-fueled power plants to either scrub out or sequester CO_2 emissions. And of equal importance, the United States should not export coal or liquidized methane. At the same time, we need increased funding efforts and tax incentives for renewable forms of energy, but only a few incentives may be necessary. We also need a plan for displaced fossil-fuel industry workers. U.S. public opinion and industrial developments are changing the face of energy production. Wind energy prices are almost on par with fossil-fueled power plants. The price of solar energy has greatly decreased in recent years and will only continue to decrease. A major drive is underway for roof-top solar-power generation that is changing the face of energy production. And well-paying jobs in renewable energy industries will far outnumber the loss of dangerous and deadly jobs in the coal industry.

In short, we need to consider everything as options for transitioning to a clean energy economy, especially given the fact that in the United States, coal-fired power plants still account for approximately 35 to 40% of our energy production. But I need to add a critical note here to my fellow environmentalists and tree huggers. There are no completely benign sources of energy. Dams destroy ecosystems and native fish populations, solar farms destroy desert tortoise habitats, current nuclear power plants generate problematic wastes, and windmills kill birds. Given human populations of Earth and our high energy needs, there will be ecological costs, and we need to choose wisely. Massive energy conservation and energy efficiency should be the first priority, since they can in fact often produce cost savings over time. Then, every relatively clean and relatively ecologically sensitive energy option should be on the table. As many have noted, however, fossil fuels will likely be needed for a decade or two to adequately supply our energy needs as we bridge the gap to a future sustainable renewable energy source. The recent developments in the United States to increase natural gas production through fracking has its detriments, but natural gas can be used as a bridge fuel to convert coal-fired power plants to natural gas and temporarily reduce our carbon footprint as we make the transition to more renewable energy sources (again, by easily half the carbon footprint). Fugitive emissions of methane gas during production are large and of concern given the persistent methane cloud being monitored over the U.S. Southwest (Borenstein 2014). CO_2 capture and sequestration for fossil fuel power plants will likely be necessary in our very near future. And new nuclear power plants may also still be required. Recent reports concerning the development of viable and low-cost nuclear fusion power plants are highly encouraging.

In the past century, through public action, social reform, and technology, we have solved global problems on a scale, some would say, of global warming. We have faced two global atmospheric disasters in recent years, acid rain and stratospheric ozone depletion, and we now have policies in place to overcome these; yet it will take decades for our final success. The scale of climate change is international and massive. We most certainly have the technology in hand to treat CO_2 from coal-fired and methane power plants—scrubbing and sequestration—as we have to develop and implement solar, wind, tidal, and many other technologies that can replace fossil fuels.

Educate yourself and your politicians. And if your politicians cannot be educated or do not believe in science, then vote them out of office.

For up-to-date news stories on climate change, visit my website at https://sites.google.com/a/whitman.edu/environmentalsuccessstories/home.

TEN

Conclusion and Transition
to a Bright Future

UNLESS someone like you cares a whole awful lot,
nothing is going to get better.
It's not.

—DR. SEUSS, *THE LORAX*

No one, not even the most optimistic tree hugger would have predicted our successes since the 1970s, when environmental degradation was commonplace. These successes set the stage for our biggest challenge yet, combating climate change. Significant cooperation, dedication, and innovation will be required to overcome the greatest and most organized case of corporate greed in the history of humans: the fossil-fuel industry. It will take all of us, and decades of hard work, to meet this challenge.

On December 12, 2015, thousands of delegates from nearly 200 countries met in Paris to finally agree on a global plan going forward. This may well be the turning point in human history where the entire world begins to collectively address human-caused climate change. I am very optimistic.

There are many excellent, science-based books and articles on what will happen if we do not act now to reduce greenhouse gas emissions and on how the fossil-fuel industry is fighting progressive change and is in denial. Examples worth consulting for more information on the nature of this problem are *Limits to Growth* by Donella Meadows, Jorgen Randers, and Dennis Meadows; *Six Degrees* by Mark Lynas; *Merchants of Doubt* by Naomi Oreskes and Erik Conway; *Deceit and Denial* by Gerald Markowitz and David Rosner; and *Storms of my Grandchildren* by James Hansen. And if we do not act in time, James Lovelock has clearly predicted what will happen next in his more recent book, *A Rough Ride to the Future*. We largely know what to do about climate change, but when will we do it? And how long after we take action will it take Earth's natural processes to reduce global atmospheric greenhouse gasses to sustainable levels? Only time will tell.

A recurring conversation at academic social gatherings is who or what event from the past would you like to personally meet or witness? Psychologists want to meet Freud, paleontologists want to see a dinosaur, Western theologians want to meet Jesus, pianists want to meet Mozart, and astronomers want to be present at the Big Bang. My personal wish would be to go forward in time, perhaps several hundred years, to see whether the human race has come to terms with Earth's limited resources, to learn whether we have overcome our most challenging environmental problem, climate change, and to hopefully witness global cooperation, global leadership, and global governance.

Keep up the fight. We only have one Earth. Again, vote for politicians who believe in science.

The closing chapter is what needs to be done now that scientists have conclusively identified the human cause of our climate-change problem. Scientists will still play an important role in renewable energy development, but now social scientists, many of whom were present at the 2015 Paris talks, need to lead the way. I have asked Dr. Kari Norgaard to explain how this vital implementation process has worked in the past and how it will work for solving climate change. Her remarks appear in the afterword.

Afterword

Imagination, Responsibility, and Climate Change

KARI MARIE NORGAARD

I appreciate Dr. Frank Dunnivant's drawing on sociological expertise to address the question of "what next?" This question is on many minds all the more following the Conference of Parties-21 (COP21) Paris climate agreement in December 2015. This is a landmark agreement in that just short of 200 nations from around the world signed on to the goal of keeping global temperatures from increasing no more than 1.5°C. Not only is the agreement a landmark event, but it is also a call to action—for it is entirely nonbinding. It will be up to all of us to somehow hold our governments accountable for reaching that goal. When it comes to environmental degradation, we have developed a solid body of knowledge from the physical and biological sciences, but all this knowledge has only partially allowed us to change course. For climate change in particular, we need to better understand the social side of why emissions are occurring. Not only does the urgency of climate change point to this need, but drawing on the expertise of the sciences as our main strategy to solve this problem has simply not worked. In the years since the dynamics of climate change have been well understood, and in the years since the International Panel of Climate Change (IPCC) was founded in 1992, the carbon dioxide concentration in the atmosphere has increased from 355 parts per million to 402 part per million in 2016.

What do we need to do to avoid those scary portions of the IPCC climate-projection graphs? With their work, the natural scientists have essentially tossed two balls into the social scientific court. One concerns the question of whether and how we can effect the change in emissions needed to avoid catastrophic climate scenarios. Now, in light of the COP21 Paris agreement, what cultural, organizational, and institutional shifts are needed to lower our carbon emissions? Can we tackle reducing our use of fossil fuels as a technical problem, or does it involve a fundamental reorganization of society? If it involves reorganization, how do we begin? The second ball calls us to understand and evaluate the impacts of changes that will come and are already taking place. In addition to figuring out how to keep climate change from getting much worse, our current emissions have already set in play a strong measure of climate change. What do we need to know just to be able to navigate the present and immediate future? To what extent will climate change lead to economic and political instability? What does a rising sea level mean for the future of cities? How is climate change exacerbating existing gender and racial inequality, and how will it do so in the near future? How for that matter, can we even begin to really imagine the reality of our present situation? Unfortunately, none of these are questions for which we have fixed answers.

As a matter of fact, climate change challenges our imaginations like never before. It seems impossible for most people to imagine the reality of what is happening to the natural world; impossible to imagine how those ecological changes are translating into social, political, and economic outcomes; and impossible to imagine how to change course. Imagination is power, especially in a time of crisis. But if we cannot even imagine the reality of what is going on, or imagine the level of change that is needed to change our course, then no forward movement will occur. Right now, in the face of climate change and environmental crises more generally, there are two specific kinds of imagination we need in both the interdisciplinary community and the public at large. We need to be able to see the relationships between human actions and their impacts on earth's biophysical system—call it an *ecological imagination*. And we need to be able to see the relationships within society that make up this environmentally damaging social structure. This second form of seeing is essentially what C. W. Mills calls a *sociological imagination*—a central concept in the field of sociology.

We now need an ecological imagination to understand the reality of our circumstances. Making visible these relationships between humans and nature has been the focus of crucial research activity in the climate arena. Atmospheric and ecological scientists have provided important descriptive evidence for the impacts of human actions on the natural world. Yet while the connection between burning fossil fuels and alteration of the climate is understood on a general level, it can still be a challenge to visualize the relationship between driving to work and the increased risk of high-intensity forest fires. This is partly because we externalize these impacts across space and time. Relationships of immediate cause and effect are masked because of processes at the atmospheric level. Emissions go into the Earth's atmosphere as a whole and the dynamics of the climate system, including processes of ocean circulation, are complex and take time to unfold. These conditions make it impossible to trace the relationship between the burning of any particular ton of carbon and the number of millimeters of sea-level rise or change in fire behavior somewhere else on the planet. But this is only part of the picture.

The scientific community has made great progress in developing our ecological imagination. Over the past four decades, atmospheric scientists have provided increasingly clear and dire assessments of alterations in the biophysical world around which human social systems are organized. When it comes to climate change, natural scientists have further laid out assessments of the necessary reductions in emissions of heat-trapping climate gases to avoid scenarios of catastrophic climate change. Yet despite these urgent warnings, human social and political responses to climate change remain wholly inadequate. We have yet to make much progress in our understanding of how to change course. For this we need for the second sociological form of imagination—the ability to see the relationships within society that make up our environmentally damaging social structure. But this more interdisciplinary, social science–based challenge has been a harder nut to crack.

From a sociological perspective, a big part of the problem lies in the fact that we are not only alienated from our ecological conditions, we are also unaware of our ability to understand the relationship between the act of getting onto an airplane and the shortened ski season. We have also become alienated from our social conditions. We see our dependence on automobiles as an individual problem—a function of poor choices rather than as a result of coordinated action by the auto industry.[1] We fail to ask how our ability to reduce our carbon footprint may be constrained by our nation's foreign policy. Essentially, we lack the ability to imagine the social structure around us. As a result, most who strive to visualize their impacts on the planet can only imagine the impact in the form of individual consumer actions (Dunlap and Brulle 2015; Shove 2010; Webb 2012). Just as it does not really make sense to think about direct linear relationships between the emissions of particular tons of carbon and specific millimeters of sea-level rise, we cannot think about individual carbon footprints in isolation from the actions of governments who set automobile fuel-efficiency standards and develop public transportation systems and the fossil fuel industry who lobbies against them.

There is an equal complexity and sophistication to the social sciences understanding of human systems, but there is less awareness from either the scientific community or the general public that it is even needed. Despite increasing calls for both interdisciplinary cooperation and the need for social science knowledge, to date little social scientific expertise has been brought to bear on the question of how we deal with present climate impacts or how to change course. With their attention to the interactive dimensions of social order between individuals, social norms, cultural systems, and political economy, sociologists are uniquely positioned to be leaders in this conversation. Yet there is only one sociologist on the IPCC, and a recent study by the International Social Science Council found that only 3% of publications in the field were by sociologists.

Climate change poses the greatest material and symbolic threat our society has encountered to date. It challenges the ethics of our daily habits and activities, it challenges our institutions and cultural norms—it even challenges our political and economic structures. Never before have we faced anything quite like this. We are in new terrain! Grappling with the science of climate change appears relatively simple by comparison, which is why Neil deGrasse Tyson recently tweeted: "In science, when human behavior enters the equation, things go nonlinear. That's why Physics is easy and Sociology is hard." Nonetheless, there are key lessons and important take-aways from existing social science understanding. Fortunately, the application of a sociological imagination and a few other sociological concepts allows us to powerfully reframe three central questions in the current interdisciplinary conversation on climate change and environmental degradation more generally: Why is climate change happening? How are we being impacted? And how might we effectively respond? Fortunately, in the course of this examination we have the possibility of a more hopeful future.

Why Is Climate Change Happening?

I mentioned earlier that many of the ideas about how we should respond to climate change emphasize individual solutions. In the words of Michael Maniates, if we are to save the world we should "ride a bike or plant a tree" (2001). This is because most discussions of why climate change is happening draw on individual assumptions regarding human behavior that neglect the role of social organization in shaping human values, choices, or understandings of the world. Tools like carbon calculators are incredibly useful for helping people to visualize that there are in fact relationships between the carbon emissions we generate when we drive or fly and impacts in the world. They help us to develop an ecological imagination. But they mislead us as well, in that they stress the concept of the individual as a meaningful unit of analysis. Sociologists will counter that climate change is due to a complex set of interactions between our economic, political, cultural, and social institutions. Individuals participate in these systems, but individual understandings, values, actions, and choices are constrained by their cultural, economic, and political contexts.

In our complex modern-world culture, technology and the economic system each shape individual experiences, values, actions, and perception. We must therefore ask how these elements of social order operate to generate such high levels of carbon emissions. Take, for example, how cultural understandings of the world are supported by technologies. Contemporary society as most of us know it rests on the cultural perception that humans are separate from nature—even that we have risen above it and are no longer vulnerable or dependent on the earth for our survival. This cultural view manifests in the lives of many people in wealthy Western nations because they can use technology to essentially control the shots. Technological transformation in our homes keep most

people warm and dry whether or not they know the source of their energy. Food systems have been restructured so that families and local
communities need not be responsible for cultivation. People who have
enough money to do so can easily move from one place to another rapidly
and over long distances using cars, trains, and airplanes. This combination
of culture (here in the form of collective beliefs) and social infrastructure
including technologies in the form of fuel sources and transportation
systems, shapes what individual people understand about the world. It
shapes what they perceive as important and whether or not they have the
sense that we are in the midst of ecological or social crises. We can thus
see how the convenience of all these activities comes with a conceptual
price. Most of us in modern Western contexts are alienated from our
ecological worlds. We no longer understand how to care for our needs
directly; social complexity and technologies buffer us from this need. And
we no longer see the consequences of our actions or perceive how serious
a collective crisis we now face.

Another key factor driving climate emissions is the structure of our
economy. Founding sociologists Karl Marx and Max Weber, both identify capitalism as the major force shaping modern societies. Many of
the leading environmental sociologists of today also point to capitalism
as central to understanding both why we are creating climate change
and why we have so far failed to effectively respond (see, e.g., Foster and
Clark 2012; Klein 2014).

But what specifically about capitalism generates such profound ecological and environment changes? One problem is that—unlike all earlier
economic systems—capitalism requires growth. While some say it may
be theoretically possible to design a steady-state market economy, in
practice, all capitalist economies that have existed to date have been
driven to grow. Others emphasize how cultural practices, institutional
infrastructure, and government entities all become part of the "treadmill
of production" under capitalism (Gould, Pellow, and Schnaiberg 2015).
Capitalism shapes the direction of technology, culture, and political structures, as well as individual values and behaviors. Take, for example, the
way that powerful industry has influenced both public understanding and
government policy on climate change in the past three decades. In the
mid-1990s awareness and concern about climate change began rapidly
rising in the United States. So great was the concern about future income

losses by the fossil fuel industry should the public become truly concerned about climate change, that individual oil companies, including Shell and Exxon, and the industry association the American Petroleum Institute developed a profoundly effective "public information campaign" designed to undermine climate science and the scientists themselves and to mislead the public (Oreskes and Conway 2010). As a result of their efforts, public understanding of the consensus on climate change was set back over two decades. At the same time, then chief of staff to the White House Council on Environmental Quality Phillip Cooney—an individual with close ties to ExxonMobil—worked inside the White House to suppress the first national-level report on the impacts of climate change in the United States. These actions led to decades of delay in government response. On a more subtle but equally powerful level, social structure shapes our cultural norms of time, cultural norms of space and conversation, can serve as "social facts" that fundamentally shape an individual's behavioral options with respect to the climate (Durkheim 2014; Norgaard 2011). We may want to ride a bike to work or school, but if there are no bike lanes, we may be taking our lives in our hands to do so. We may want to lower our carbon footprint, but if our energy grid is supplied by coal, any efforts to save household energy will go only so far. We may want to have ethical work options, but if we live in poverty, we may not have much choice. We may want to use our engineering degree to improve the world, but if the well-paying jobs are all in extractive mining it may be a hard choice to make.

How Are We Being Impacted?

The social impacts of environmental degradation generally and climate change in particular are neither uniform nor entirely predicable in a strict linear sense. Although the changing climate will eventually impact everyone, it has already begun to precipitate the most extensive and violent impacts to date against the poor, women, and people of color here in the United States and around the globe. As illustrated by the aftermath of Hurricane Katrina or the need to relocate the inhabitants of disappearing island nations in the South Pacific, the impacts of extreme weather events associated with climate change are disproportionately borne along

the lines of gender, race, and class, both within and across national boundaries. As a result, the ecological changes that undermine social activities and infrastructure simultaneously reproduce gender, racial, and class
inequalities in complex ways depending on social context. Yet racism and
class spatial segregation whether in the form of gated communities or
just people living in separate parts of town make these impacts and realities less visible to the largely white and male decision makers in county,
state, and federal government. And to the extent that global inequalities
mask the extent and severity of changing ecological conditions, inequality
itself can be seen as a contributing factor to climate change.

Climate effects manifest differently around the world according to
geography, as well as the particular local conditions within different each
social context. Social disruptions such as extreme heat events will cause
more deaths among those with fewer economic resources. Women are
more vulnerable in disaster situations, not only because of increased poverty, but also because they are more likely to be caring for young children
or the elderly. Unequal impacts of climate change can be acute, during
crisis events such as flooding or hurricanes (Fothergill and Peek 2004), or
ongoing over long periods of time, as with extended droughts. Thus, on
the one hand, women working in subsistence activities are particularly
vulnerable because of long-term changing flood and climate regimes, while
African American women post-Katrina have faced more difficulty securing loans to rebuild businesses and homes than their white male counter-
parts (Bullard and Wright 2009).

Many Americans think of climate change and inequality in terms of
impacts to distant places such as Bangladesh. But not all of those who are
most impacted live in the Global South. Fifty percent of the citizens of the
United States live within 50 miles of a coast. Our coastal cities are and will
be greatly impacted. Age is a particularly important dimension of vulnerability in that climate impacts will be carried forward into the future. For
this reason, young people today are calling for "the right to a future," and
"intergenerational climate justice." There is now a strong and rapidly
growing youth justice movement in the United States and worldwide.[2]
And race, gender, and class radically shape experiences of communities
within nations of the Global North as well. For example, indigenous communities in the U.S. and Canadian Arctic are on the front lines of direct
impacts from climate change (Arctic Climate Impact Assessment 2004).

Arctic indigenous communities face more direct impacts because of combinations of ecological and social factors including their location in the Arctic, where changes are occurring more rapidly, and because of more intact cultural ties to land such that people retain subsistence activities and culture and social structure are organized around conditions of the natural world. Within indigenous communities, women may face added burdens in the form of increased reproductive labor—maintaining family and community cohesion, as when entire island communities are displaced by storm surge and relocated to the mainland (Tandon 2011). Thus, the particular effects of climate change reorganize gender and racial community relations in unique and often complex ways. The groups of people who bear a greater impact also contribute the least to carbon emissions, making climate change clearly an issue of global environmental justice.

Economic and social inequality coupled with the reorganization of time and space under globalization can also serve as a contributing cause of climate change. The social distance engendered by inequality masks collective understanding and awareness of the immediacy of climate impacts. Anthony Giddens's concept of "time-space distanciation" describes how globalization creates "the intensification of worldwide social relations which link distant localities in such a way that local happenings are shaped by events occurring many miles away and vice versa. . . . In the modern era, the level of time-space distanciation is much higher than in any previous period, and the relations between local and distant social forms and events become correspondingly 'stretched'" (1991:64).

As a result of both inequality and distanciation, climate change appears distant or even invisible to those with the time, energy, cultural capital, and political clout to generate moral outrage in a variety of ways. Indeed we see a pattern whereby people in nations with higher carbon emissions and greater political and economic power are also less likely to be concerned about climate change. Social inequality becomes part of the distance that makes climate change appear as no more than background noise for these more privileged social actors. Social inequality is growing rapidly in the United States, to the point that we are losing the middle class. It is harder and harder for middle-class families to send their children to college and more and more young people leave college with enormous debt, a fact that itself is responsible for reshaping the opportunities for young people after graduation.

How Might We Effectively Respond?

For over 20 years atmospheric scientists have provided increasingly clear and dire assessments of alteration in the biophysical world around which human social systems are organized. Natural scientists have further laid out assessments of the necessary reductions in emissions of heat-trapping climate gases to avoid scenarios of catastrophic climate change. Yet, paradoxically, we see remarkably little public response. How can we change course and respond effectively?

My own work on this question concerns the notion of climate denial. This is not literal denial of the so-called climate skeptic movement in the sense of "climate change is not happening," but rather it is what Stanley Cohen (2013) calls "implicatory denial," or the way in which large majorities of people in Western democracies are aware of climate change and believe in it but are able to act as though it is not happening. This is the notion of denial as the elephant in the room.

Both my research in Norway and my follow-up work in the United States describes how for many people thinking seriously about climate change evokes a series of troubling emotions (Norgaard 2011, 2014). There is fear about a future with more heat waves, more droughts, and increased storm intensity. There is fear that our present political and economic structures are unable to effectively respond. And for many there is guilt, since Americans are among the main contributors to global climate emissions. Finally, many people described a sense of not knowing what to do in the face of climate change. Ultimately, reducing global climate emissions sufficiently is beyond the level of individual action. But national and international efforts have also been unsuccessful. Awareness of this generates for many a feeling of helplessness. These are unpleasant emotions, and it turns out we have many ways to normalize them, to look the other way, to not think about the unthinkable. One implication of socially organized denial of climate change is that as individuals we must struggle to imagine the reality of our current situation.

This kind of denial poses a serious threat to democracy, for if we do not act in a crisis we cannot resolve it. This kind of denial in the form of apathy makes our present predicament appear all the more disturbing in a kind of circular manner, for it appears that there is a major crisis but nobody seems to even care. If climate change poses a new challenge for individuals to live morally coherent and responsible lives, it simultaneously poses a new challenge for democracy because of the widespread problem of public climate apathy or climate denial. If we all sit like proverbial frogs in boiling water then our democracy is surely not well. As things now stand, the human capacity for denial in the face of problems that feel too large to tackle threatens to erode the critical democratic role of the so-called public sphere at a time when we would seem to need it more than ever. Without sensing a genuine reason to engage, individuals withdraw from the political as a self-protective response. Yet for democracy to work, it needs engaged citizens thinking and caring and acting ethically on behalf of the larger society.

In the midst of denial and public apathy there has been a renewed effort at science communication. The assumption here is that people just don't get it, and lack of information is a key limiting factor in our failed public response. Science communication is now in fact a field in itself. This is a leading area of publishing activity in the interdisciplinary quest for responding to climate change. Sessions on this topic are organized at all the major scientific conferences from the American Geophysical Union to the Society for Conservation Biology. As powerful and necessary as most climate communications strategies may be, they have several limitations from a sociological perspective. A key limitation is that these communications are aimed at individuals—but individuals cannot solve climate change on their own, and they know it. Emissions reductions at the scale we need must be achieved at the collective level. For Giddens, a central part of the problem here is that "we have no politics of climate change" (2009:4).

Climate change challenges our existing social order like no prior social or ecological problem, so developing that politics is no straightforward matter. Climate change is disturbing not only because of the social and ecological collapse that is predicted, but also because of the apparent possibility that our existing political structures are not up to the task of dealing with climate change. Unless we can refashion our political and economic systems, we are trapped. We need to see the reality of our lives and break through the absurdity of the double life. However, we face a circular situation. When political leadership is captured by the fossil fuels lobby and fails to send a clear signal about climate change, it contributes to a sense of hopelessness. When people do not see others standing up around them, it contributes to cultural norms of complacency. And when we hear around us only conversations about riding bikes or changing light bulbs, it becomes difficult to imagine how real change could occur. The fact that many people perceive no viable political options is a central part of why they are not responding. Individual apathy is a rational response if there is nowhere to turn and no hope that investing our energy will pay off. We become self-protective and retreat into smaller and smaller social worlds. In order to mobilize people, we need real solutions. And here we see a second limitation to many efforts. Because people find real change beyond their imagination, politically unfeasible, or not in their economic and political interests, they look to solutions that tinker with the system instead. So we need to have a real conversation about what can be done and how each of us can be a part of creating a healthy and livable future.

There Are Many Things We Can Do

If climate apathy is a cycle held in place by individual fear and silence, complicit cultural norms, and a state logic based on fossil fuel extraction and economic profit at any politically acceptable cost, then this cycle can be interrupted at multiple points. There are many things we can do, and now is the time. While the COP21 agreement in Paris in December 2015 is monumental with regard to the number of nations that came to the table and agreed to make change, it is also entirely nonbinding. This means that it is up to communities on the ground to be sure that our political leaders figure out how to keep fossil fuels in the ground. Key strategic possibilities exist in any political struggle. In our present times of rapid social change, such strategic moments will continue to emerge, and we can be ready for them. More generally, individual people can get involved in the many ongoing local, regional, and national political efforts from those organized by 350.org or Citizens Climate Lobby to events in churches, on college and university campuses, divestment movements, and more. Indeed there are so many things that people *are* doing every day.[3] As I write this, students at my own campus at University of Oregon are sitting in at the university administration office for the third week in a row as part of the campaign to divest from fossil fuels. Social theorists like Hannah Arendt (Arendt 1970) remind us of the importance of power from below—even talking about climate change with family and friends is an important way to break the present cultural silence. Talking to our peers and the people around us is part of the development of not only a sociological imagination, but also a political one. Arendt describes how as people get together and speak in small groups, an important kind of meaning making power takes place that is an essential element of democracy.

Although in isolation they are not enough, local efforts to make climate change visible in one's community, to plan for coming changes in the water supply, to change energy usage, and to reduce emissions at the county and regional levels that are based on existing community ties and sense of place and identity may provide a key for breaking through climate denial from the ground up. There is already a global movement building for communities to uncover how climate change is manifesting in their local contexts. Local political renewal cannot be enough on its own. But it may be the important next step for individuals in breaking through the absurdity of the double life and for renewing the democratic process. As each of us thinks about what is happening in our own place and how we want to respond, we will begin to see why the facts of climate change matter in our own lives and can develop a sociological imagination at the same time as we reconnect the rifts in time and space that have made climate change a distant issue. Working together may over time create the supportive community that is a necessary (though not sufficient) condition for each of us to face large fears about the future and engage in large-scale social change. Facing climate change will not be easy. But it is worth trying.

Our Society Is Not Static

I hear many conversations these days about how we are inevitably on the wrong path, or that our fate with respect to climate change is predetermined, either because we are hardwired to not understand the science, because we are hardwired to act in a narrow-minded individual self-interest, or because of our inevitably compromised political system. True, the stakes are extremely high, and our record to date is not good. Yet while we are surely in a crisis, it is not the case that there is nothing we can do. Some of these theories come from discouraged public intellectuals, others from members of the scientific community. From a sociological perspective, the first thing to realize is that our society is not static. Furthermore, as a sociologist I advise you to consider who benefits from these narratives. Narratives that focus on individual responsibility let corporate actors and big politicians off the hook. Similarly nothing quells public outrage more quickly than the notion that there is nothing one can do. By contrast, it is clear that we are living in times of rapid social change. Not only is the biophysical world around which our society is organized sending us more large-scale climactic events, we are currently in what social movement theorists call "protest cycles"—many larger social movements are emerging, a fact that can radically change the political terrain in a short time. Our society is certainly not static. There are plenty of people who are working on climate change, in fact more people every hour, and there are plenty of opportunities to engage.

Climate Change as an Opportunity

The people I work with on the Klamath River talk about climate change as an opportunity. In her book *This Changes Everything: Capitalism vs. the Climate*, Naomi Klein takes this angle as well. How could this be so? Many aspects of our present economic and political structure are less than ideal—from the falling middle class, to rising student loan debt, and declining public infrastructure as evidenced by happenings in New Orleans and Flint, Michigan. We have profound levels of environmental contamination, species decline, social inequality, and hunger and many forms of human suffering. The fact that climate change is a profound challenge to our present world can be an opportunity to reexamine the failing assumptions of our current economic and political order. It can be an opportunity to do better. We have been undermining our ecological base for some time. Climate change comes as a wake-up call for many.

We Have a Responsibility

Lastly, I personally believe that each of us has a responsibility to engage. We do not get to give up. For over a decade I have been doing collaborative research on behalf of the Karuk Tribe in California. These people are working hard to restore their river and forest ecosystem in the face of dams that block salmon passage, failed forest policies, unratified treaties, and now climate change. I have learned a great deal from my tribal colleagues over the past dozen years. Perhaps the thing that stands out the most is the profound sense of responsibility that people feel to care for one another and the physical world. In the words of my Karuk colleague and friend Ron Reed, "We have a responsibility to take care of the fish, to take care of the forest, and to our families, to our Elders." This sense of responsibility to care for one's fellow beings stands out in sharp contrast to the narrative of "rights" that is so prevalent elsewhere in the West. For tens of thousands of years across North America the relationships between people and places were governed by an ethic of responsibility. If people were to hunt or fish they were responsible for tending to the land in ways that created better habitats for those species to flourish. Both people and the Earth thrived under these relationships. The settlers that came from around the world to colonize and develop what we now know as the United States of America have achieved many things, but when it comes to our relationship with the environment, one hears more often a discourse of land rights, water rights, and rights to develop natural resources than responsibilities to care for the land, one another, and future generations. I believe it is time to return to an ethic of responsibility.

Most of you reading this book are in a situation of relative privilege. Your prosperity (however moderate or great) came about in part through the use of fossil-fuel–driven economic activity. You have a fair bit of education. We are in a position to make a difference and have a responsibility to do our best. Yet this responsibility does not need to be a burden. Engaging in forms of action for our world, however large or small, can bring empowerment, friendships, and a sense of meaning in the world, and these can be fun. Maybe one of you reading this will develop an addition to the carbon calculator that helps people visualize a sociological as well as the ecological imagination. . . .

Notes

1. A great documentary on this is *Who Killed The Electric Car?* (Dir. Chris Paine; Electric Entertainment, 2006).

2. Theo LeQuesne, "Post-Paris Activism: How to Build an Effective Global Struggle to Tackle Climate Change," *Lacuna Magazine* (February 2016), http://lacuna .org.uk/insider/post-paris-activism-how-to-build-an-effective-global-struggle-to -tackle-climate-change.

3. Among many exciting things is the development of the Leap Manifesto in Canada: https://leapmanifesto.org/en/the-leap-manifesto.

Bibliography

Arctic Climate Impact Assessment. *Impacts of a Warming Arctic.* Cambridge: Cambridge University Press, 2004.

Arendt, H. *On Revolution.* New York: Viking, 1970.

Bullard, R. D., and B. Wright. *Race, Place, and Environmental Justice After Hurricane Katrina: Struggles to Reclaim, Rebuild, and Revitalize New Orleans and the Gulf Coast.* Boulder, CO: Westview Press, 2009.

Cohen, S. *States of Denial: Knowing About Atrocities and Suffering.* New York: Wiley, 2013.

Dunlap, R., and R. Brulle, eds. *Sociological Perspectives on Climate Change* (Report of the ASA Task Force on Sociology and Global Climate Change). New York: Oxford University Press, 2015.

Durkheim, É. *The Rules of Sociological Method: And Selected Texts on Sociology and Its Method.* New York: Simon and Schuster, 2014.

Foster, J. B., and B. Clark. "The Planetary Emergency." *Monthly Review* 2012;64(7):1–25.

Fothergill, A., and L. A. Peek. "Poverty and Disasters in the United States: A Review of Recent Sociological Findings." *Natural Hazards* 2004;32(1):89–110.

Giddens, A. *Modernity and Self-Identity: Self and Society in the Late Modern Age.* Stanford, CA: Stanford University Press, 1991.

Giddens, A. *The Politics of Climate Change.* Cambridge: Polity, 2009.

Gould, K. A., D. N. Pellow, and A. Schnaiberg. *Treadmill of Production: Injustice and Unsustainability in the Global Economy.* New York: Routledge, 2015.

Klein, N. *This Changes Everything: Capitalism vs. the Climate.* New York: Simon and Schuster, 2014.

Maniates, M. "Individualization: Buy a Bike, Plant a Tree, Save the World?" In *Confronting Consumption* (pp. 43–66), edited by T. Princeton, M. Maniates, and K. Conca. Cambridge, MA: MIT Press, 2002.

Mills, C. W. *The Sociological Imagination.* New York: Oxford University Press, 1959.

Norgaard, K. M. *Living in Denial: Climate Change, Emotions and Everyday Life.* Cambridge, MA: MIT Press, 2011.

Norgaard, K. M. "Normalizing the Unthinkable: Climate Denial and Everyday Life." In *Twenty Lessons in Environmental Sociology*, 2nd ed., edited by K. Gould and T. Lewis. Oxford: Oxford University Press, 2014.

Oreskes, N., and E. M. Conway. *Merchants of Doubt.* New York: Bloomsbury Press, 2010.

Shove, E. "Beyond the ABC: Climate Change Policy and Theories of Social Change." *Environment and Planning A* 2010;42:1273–1285.

Tandon, N. "Climate Change: Beyond Coping. Women Smallholder Farmers in Tajikistan." *Oxfam Policy and Practice: Agriculture, Food and Land* 2011;11(3):114–165.

Webb, J. "Climate Change and Sociology: The Chimera of Behavior Change Techniques." *Sociology* 2012;46(1):109–125.

Bibliography

One of the most challenging aspects of writing this book was the selection of references to include. Since 2004, when I started collecting items for this book, I have read a few thousand scientific articles, science summaries, news reports, and books. Given that this is a popular science book and not especially for scientists who are already familiar with these topics, I have chosen to include only readings that I feel are accessible to lay readers. But I have also included some of the essential scientific articles on major subjects. One of my main sources of new briefs, science, industry, and government summaries over the past decade has been from a chemical trade magazine/journal called *Chemical Engineering and News* (C&E News). I have included some of these more relevant articles from C&E News, and many of these contain the reference that will lead you to the scientific literature.

For up-to-date news stories on environmental issues please see my website, which is updated daily, at https://sites.google.com/a/whitman.edu/environmental successstories.

Introduction

Further Reading

Schnoor, J. L. "Top 10 Environmental Success Stories." *Environmental Science and Technology* 2004a;September 1:319A.
Schnoor, J. L. "Top 10 Stupid Environmental Policies." *Environmental Science and Technology* 2004b;July 1:239A.

1. Securing Safe, Inexpensive Drinking Water

Notes

Ayres, K. "NM Starts 2014 with Severe Drought." KRQE News 13, March 22, 2014.

Cape Fear River Assembly. "Water Quotes." http://cfra-nc.org/water-quotes (accessed August 2, 2016).

Pindar, A. Z. http://www.azquotes.com/quote/539201 (accessed August 2, 2016).

UN Millennium Development Goals. http://www.un.org/millenniumgoals (accessed June 3, 2014).

UN Sustainable Development Goals. https://sustainabledevelopment.un.org/index .php?menu=1300 (accessed August 3, 2016).

World Bank. *Mini-Atlas of Millennium Development Goals: Building a Better World.* Washington, DC: Author, 2005.

Further Reading

Amato, I. "Making Troubled Waters Potable." *Chemical and Engineering News* 2009;April 17:39–40.

American Water Works Association. http://www.awwa.org (accessed June 3, 2014).

Arnaud, C. H. "Figuring Out Fracking Wastewater." *Chemical Engineering and News* 2015;March 16:9–14.

Baker, M. N. *The Quest for Pure Water,* Vol. 1, 2nd ed. Denver, CO: American Water Works Association, 1981.

Baum, R. M. "Water, Water Everywhere." *Chemical and Engineering News* 2011; February 21:3.

Blue the Water Issue: A Special Report, 4th Annual Environmental Issue, Vol. 6, no. 1, 2003.

CAS. "CAS Registry and CAS Registry Number FAQs." http://www.cas.org/content /chemical-substances/faqs (accessed June 3, 2014).

Das, B. "Double Trouble with Arsenic." *Chemical and Engineering News* 2015; August 3:38–39.

Devarajan, S., M. J. Miller, and E. V. Swanson. *Goals for Development: History, Prospects, and Costs.* Washington, DC: World Bank, April 2002. Policy Research Working Paper 2819, http://econ.worldbank.org/external/default/main?pagePK =64165259&theSitePK=469372&piPK=64165421&menuPK=64166093 &entityID=000094946_02041804272578.

EPA. "Clean Water Act." http://www2.epa.gov/laws-regulations/summary-clean-water -act (accessed June 3, 2014).

EPA. *The History of Drinking Water Treatment.* Washington, DC: Office of Water, EPA-816-F-00-006, February 2000.

EPA. "Water Topics." http://water.epa.gov (accessed June 3, 2014).

Erickson, B. E. "Cleaning the Chesapeake." *Chemical and Engineering News* 2011; December 12:10–14.

Girard, J. E. *Principles of Environmental Chemistry* (pp. 182–209). Sudbury, MA: Jones and Bartlett, 2010.

Halliday, S. *Water: A Turbulent History.* Gloucestershire, UK: Sutton, 2004.

Hawley, C., Science-Associated Press. "U.N. Says Globe Drying Up at Fast Pace," June 15, 2004. http://www.worldrevolution.org/article/1465 (accessed June 2, 2014).

Hogue, C. "Will Congress Clear the Water?" *Chemical and Engineering News* 2006;June 26:7.

Johnson, J. "BP's Ever-Growing Oil Spill." *Chemical and Engineering News* 2010;June 14:15–24.

Jones, M. "Not Again: Flint and the Consequences of Following the Law Instead of Its Intent." *Industry Voices*, January 29, 2016. https://communities.acs.org /community/society/industry-voices/blog/2016/01/29/not-again (accessed August 16, 2016).

Kandel, R. *Water from Heaven: The Story of Water from The Big Bang to the Rise of Civilization, and Beyond.* New York: Columbia University Press, 2003.

Kemsley, J. "Reckoning with Oil Spills." *Chemical and Engineering News* 2015; July 13:8–12.

Kemsley J. "Treating Sewage for Drinking Water." *Chemical and Engineering News* 2008;January 28:71–73.

Levine, A. D., and T. Asano. "Recovering Sustainable Water from Wastewater." *Environmental Science and Technology* 2004;June 1:201–208A.

McLoughlin, P. "Scientists Say Risk of Water Wars Rising." *Environmental News Network*, August 20, 2004. http://www.enn.com/top_stories/article/11097 (accessed June 2, 2014).

Morrison, J. "Clearing Up Muddy Water." *Chemical Engineering and News* 2014; December 1:25.

Morrison, J. "Fracking Study Gives Mixed Results." *Chemical Engineering and News* 2015;June 15:23.

National Environmental Services Center. www.nesc.wvu.edu (accessed June 3, 2014).

Pacific Institute. "The World's Water." http://worldwater.org (accessed August 10, 2016).

Patrick, R. *Surface Water Quality: Have the Laws Been Successful?* Princeton, NJ: Princeton University Press, 1992.

Pearce, F. "The Parched Planet." *New Scientist* 2006;February 25.

Reisch, M. S. "Dehydrated Water, Disaster Relief Drinks." *Chemical and Engineering News* 2011;February 21:41.

Roddick, A. *Troubled Water: Saints, Sinners, Truths and Lies About the Global Water Crisis.* West Sussex, UK: Roddick, 2004.

Schnoor, J. L. "No Water, No Future." *Environmental Science and Technology* 2004;August 1:279A.

Scott, A. "Running Dry." *Chemical and Engineering News* 2013;July 22:10–15.

Short, P. L. "Keeping It Clean." *Chemical and Engineering News* 2007;April 23:13–20.

TDSmeter.com. "History of Water Quality, Purification, and Filtration." http://www.tdsmeter.com/education?id=0002 (accessed June 3, 2014).

Torrice, M. "How Lead Ended Up in Flint's Water." *Chemical and Engineering News* 2016;Feburary 15:26–29.

Tremblay, J. F. "Cash Flows from China's Water." *Chemical and Engineering News* 2009;May 11:18–21.

Tremblay, J. F. "China Tightens the Screw." *Chemical and Engineering News* 2016;January 4:11–13.

Tullo, A. "Obscure Chemical Taints Water Supply." *Chemical and Engineering News* 2014;February 17:10–15.

USA Today. "Arsenic in Drinking Water Seen as Threat," August 30, 2007. http://usatoday30.usatoday.com/news/world/2007-08-30-553404631_x.htm (accessed August 10, 2016).

Water World Council. http://www.worldwatercouncil.org/index.php?id=1 (accessed August 10, 2016).

Water World Forum. http://www.worldwaterforum6.org/en (accessed August 10, 2016).

World's Freshwater Resources. http://worldwater.org (accessed June 2, 2014).

Yabroff, J. "$3 Gadget Produces Safe Drinking Water." *Newsweek*, March 13, 2010. http://www.newsweek.com/3-gadget-produces-safe-drinking-water-102557 (accessed August 10, 2016).

Yabroff, J. "Water for the World." *Newsweek*, June 18, 2007.

2. Effective Treatment of Our Wastewaters

Notes

Coleridge, S. T. Poemhunter.com. http://www.poemhunter.com/poem/cologne (accessed August 4, 2016).

Koplin, D., E. T. Furlong, M. T. Meyer, E. M. Thurman, S. D. Zaugg, L. B. Barber, H. T. Buxton. "Pharmaceuticals, Hormones, and Other Organic Wastewater Contaminants in U.S. Streams, 1999–2000: A National Reconnaissance." *Environmental Science and Technology* 2002;March 13:1202–1211.

Meadows, D., J. Randers, and D. Meadows. *Limits to Growth: The 30th-Year Update* (figure 3-22, p. 112). White River Junction, VT: Chelsea Green, 2004.

Rosen, C. M. "'Knowing' Industrial Pollution: Nuisance Law and the Power of Tradition in a Time of Rapid Economic Change, 1840–1865." *Environmental History* 2003;October:63–595.

Schuster, P. F., D. P. Krabbenhoft, D. L. Naftz, L. D. Cecil, M. L. Olson, J. F. Dewild, D. D. Susong, J. R. Green, and M. L. Abbott. "Atmospheric Mercury Deposition During the Last 270 Years: A Glacial Ice Core Record of Natural and Anthropogenic Sources." *Environmental Science and Technology* 2002;April 24:2303–2310.

Wikipedia. "John Snow." https://en.wikipedia.org/wiki/John_Snow (accessed August 4, 2016).

Wikipedia. "Louis Pasteur." https://en.wikipedia.org/wiki/Louis_Pasteur (accessed August 4, 2016).

Wikipedia. "Robert Koch." https://en.wikipedia.org/wiki/Robert_Koch (accessed August 4, 2016).

World Bank. *Mini-Atlas of Millennium Development Goals: Building a Better World.* Washington, DC: Author, 2005.

Further Reading

American Water Works Association. http://www.awwa.org (accessed June 3, 2014).

Anderson, H., H. Siegrist, B. Hallings-Sorensen, and T. A. Ternes. "Fate of Estrogens in a Municipal Sewage Treatment Plant." *Environmental Science and Technology* 2003;July 26:4021–4026.

Arnaud, C. H. "Figuring Out Fracking Wastewater." *Chemical Engineering and News* 2015;March 16:9–14.

Ashley, K., D. Cordell, and D. Mavinic. "A Brief History of Phosphorus: From the Philosopher's Stone to Nutrient Recovery and Reuse." *Chemosphere* 2011;84:737–746.

Baird, C. *Environmental Chemistry* (pp. 287–322). New York: Freeman, 1995.

Blumenthal, L. "As Oceans Fall Ill, Washington Bureaucrats Squabble." *McClatchy D.C.,* November 8, 2009. http://www.mcclatchydc.com/2009/11/08/78489/as-oceans-fall-ill-washington.html (accessed November 8, 2009).

Blumenthal, L. "Growing Low-Oxygen Zones in Oceans Worry Scientists." *McClatchy D.C.,* March 7, 2010. http://www.mcclatchydc.com/2010/03/07/89918/growing-low-oxygen-zones-in-oceans.html (accessed May 8, 2010).

Boxall, A. B., D. W. Kolpin, A. H. Sorensen, and J. Tolls. "Are Veterinary Medicines Causing Environmental Risks?" *Environmental Science and Technology* 2003; August 1:286-294A.

Centers for Disease Control and Prevention (CDC). "Water." http://www.cdc.gov/healthywater (accessed June 4, 2014).

Chemical and Engineering News. "Don't Blame the Pill." 2010;November 1:6.

Chio, C. "The Big Story. American's New Love." Associated Press, March 11, 2013. http://bigstory.ap.org/article/americas-new-love-water (accessed February 12, 2013).

Cooney, C. M. "UNEP's Voluntary Mercury Approach." *Environmental Science and Technology* 2005;June 1:237A.

Dastoor, A. P., and Y. Larocque. "Global Circulation of Atmospheric Mercury: A Modeling Study." *Atmospheric Environment* 2004;38:147–161.

Devarajan, S., M. J. Miller, and E. V. Swanson. *Goals for Development: History, Prospects, and Costs.* Policy Research Working Paper 2819. Washington, DC: World Bank, April 2002. http://econ.worldbank.org/external/default/main?pagePK =64165259&theSitePK=469372&piPK=64165421&menuPK=64166093 &entityID=000094946_02041804272578.

Donn, J. "AP IMPACT: Tons of Released Drugs Taint US Water." *U.S. News,* April 19, 2009. http://www.usnews.com/science/articles/2009/04/19/tons-of-released-drugs -taint-us-water (accessed April 20, 2009).

Donn, J. "Study: Fish in Drug-Tainted Water Suffer Reaction." *Yahoo News,* February 14, 2013. http://news.yahoo.com/study-fish-drug-tainted-water-suffer-reaction -190727109—finance.html (accessed February 14, 2013).

Erickson, B. E. "Dead Zone: Smaller, but More Severe." *Chemical and Engineering News* 2009;August 3:22.

Everts, S. "Drugs in the Environment." *Chemical and Engineering News* 2010; March 29:23–24.

Girard, J. E. *Principles of Environmental Chemistry* (pp. 214–246). Sudbury, MA: Jones and Bartlett, 2010.

Halford, B. "Side Effects." *Chemical and Engineering News* 2008;February 25.

Hileman, B. "Cleaning Up the Great Lakes." *Chemical and Engineering News* 2002;June 17:23–24.

Hogue, C. "Dead Zone." *Chemical and Engineering News* 2006;October 2:40–42.

Hogue, C. "'Dead Zones' Mount Along U.S. Coasts." *Chemical and Engineering News* 2010;September 13:18.

Hogue, C. "Gulf of Mexico: Dead Zone Peaks at Near-Record Size." *Chemical and Engineering News* 2010;August 9:9.

Hogue, C. "Of Algae and Dead Zones." *Chemical and Engineering News* 2009; October 5:33–34.

Hsu, J. "Ocean Garbage Patch Still a Mystery." *Live Science,* August 19, 2010. http:// www.livescience.com/10027-ocean-garbage-patch-mystery.html (accessed June 2, 2014).

Kauffman, G. B. "Drugs in the Environment." *Chemical and Engineering News* 2008;December 15.

Kolodzief, E. P., Harter, T., and D. L. Sedlak. "Dairy Wastewater, Aquaculture, and Spawning Fish as Sources of Steroid Hormones in the Aquatic Environment." *Environmental Science and Technology* 2004;August 21:6377–6383.

Loffler, D., J. Rombke, M. Meller, and T. A. Ternes. "Environmental Fate of Pharmaceuticals in Water/Sediment Systems." *Environmental Science and Technology* 2005;June 9:5209–5218.

Mason, M. "World's Highest Drug Levels Entering India Stream." *Jakarta Post,* January 26, 2009. http://www.thejakartapost.com/news/2009/01/26/world039s -highest-drug-levels-entering-india-stream.html (accessed June 2, 2014).

McCoy, M. "Goodbye, Phosphates." *Chemical and Engineering News* 2011;January 24:12–17.

Moissse, K. "Arsenic Flowing Into NC River Hit by Coal Ash Disaster." *Good Morning America/Yahoo,* February 19, 2014. http://gma.yahoo.com/blogs/abc-blogs/arsenic -flowing-nc-river-hit-coal-ash-disaster-175254843—abc-news-health.html (accessed June 4, 2014).

Pacific Institute. *The World's Water.* http://worldwater.org (accessed June 2, 2014).

Pathak, B. "The History of Public Toilets." https://www.plumbingsupply.com/toilet historyindia.html (accessed February 2013).

Pelley, H. "Estrogen Knocks Out Fish in Whole-Lake Experiment." *Environmental Science and Technology.* 2003;September 1:313–314A.

Pickering, A. D., and J. P. Sumpter. "Comprehending Endocrine Disrupters in Aquatic Environments." *Environmental Science and Technology* 2003;September 1:331–336A.

Proceedings of the UN Economic and Social Council 29th Session. Geneva, November 11–29, 2002. E/C.12/2002/11. http://www1.umn.edu/humanrts/gencomm /escgencom15.htm.

Renner, R. "Small Fish for Healthy Babies." *Environmental Science and Technology* 2004;June 15:262–263A.

Rouh, A. M. "Whereto, Mercury?" *Chemical and Engineering News* 2001;September 24:35–38.

Royte, E. "A Tall, Cool Drink of . . . Sewage?" *New York Times,* August 8, 2008. http:// www.nytimes.com/2008/08/10/magazine/10wastewater-t.html?_r=2&oref =slogin& (accessed June 2, 2014).

Sanudo, S. A., and G. A. Gill. "Impact of the Clean Water Act on the Levels of Toxic Metals in Urban Estuaries: The Hudson River Estuary Revisited." *Environmental Science and Technology* 1999;August 12:3477–3481.

Schladweiler, J. C. "Tracking Down the Roots of Our Sanitary Sewers." http://www .sewerhistory.org/chronos/roots.htm (accessed February 2013).

Scott, A. "Cleaning Up Drugs in Wastewater." *Chemical and Engineering News* 2015;August 3:24–25.

Sierra Club. "Coal Ash Sites in the U.S." http://www.sierraclub.org/coal/coalash /map.aspx (accessed June 3, 2014).

Simmon, R. "What Causes Dead Zones." *New Scientist* 2012;September 25. http://www .scientificamerican.com/article/ocean-dead-zones (accessed August 10, 2016).

SourceWatch. http://www.sourcewatch.org/index.php/Existing_U.S._Coal_Plants (accessed December 15, 2014).

Sumpter, J. P., and A. C. Johnson. "Lessons from Endocrine Disruption and Their Application to Other Issues Concerning Trace Organics in the Aquatic Environment." *Environmental Science and Technology* 2005;May 11:4321–4329.

Sundberg, E. "7–11 and Wegmans Bottled Water Recalled Due to E. Coli Concerns." *Yahoo Health*, June 22, 2015. https://www.yahoo.com/health/7-eleven-and-weg mans-bottled-water-recalled-due-to-122181523142.html (accessed August 31, 2015).

Ternes, T. A., A. Joss, and H. Siefrist. "Scrutinizing Pharmaceuticals and Personal Care Products in Wastewater Treatment." *Environmental Science and Technology* 2004;October 15:393–398A.

Thacker, P. D. "The Brain Is Defenseless Against Mercury." *Environmental Science and Technology* 2005;June 1:234A.

Tremblay, J. F. "Cash Flows from China's Water." *Chemical and Engineering News* 2009;May 11:18–21.

Tremblay, J. F. "China Changes Gear." *Chemical and Engineering News* 2002;November 10:51–53.

Tremblay, J. F. "The Dark Side of Indian Drug-Making." *Chemical and Engineering News* 2011;January 3:13–14.

Turner, R. E., N. N. Rabalais, and D. Justic. "Gulf of Mexico Hypoxia: Alternate States and a Legacy." *Environmental Science and Technology* 2008;February 20:2323–2327.

UN Millennium Development Goals. http://www.un.org/millenniumgoals (accessed June 3, 2014).

Wikipedia. "2014 Elk River Chemical Spill." http://en.wikipedia.org/wiki/2014_Elk _River_chemical_spill (accessed June 3, 2014).

Yan, H., and B. Blanchard. "Garbage Islands Threaten China's Three Gorges Dam." Reuters, August 2, 2010. http://www.reuters.com/article/2010/08/02/us-china -threegorges-idUSTRE6710SH20100802 (accessed June 2, 2014).

Young, L. "Vancouver's Toxic Sewage Dumped in Ocean." *The Thunderbird*, May 4, 2008. http://thethunderbird.ca/2008/05/04/vancouvers-toxic-sewage-dumped-in -ocean (accessed June 2, 2014).

3. The Removal of Anthropogenic Lead, and Soon Mercury, from Our Environment

Notes

Altman, S. http://www.brainyquote.com/quotes/authors/s/sidney_altman.html (accessed August 4, 2016).

EPA. "Advisories and Technical Resources for Fish and Shellfish Consumption." http://water.epa.gov/scitech/swguidance/fishshellfish/fishadvisories/index.cfm (accessed June 4, 2014).

Kittman, J. L. "The Secret History of Lead: Special Report." *The Nation*, March 20, 2000. http://www.thenation.com/article/secret-history-lead (accessed June 3, 2014).

Schuster, P. F., D. P. Krabbenhoft, D. L. Naftz, L. D. Cecil, M. L. Olson, J. F. Dewild, D. D. Susong, J. R. Green, and M. L. Abbott. "Atmospheric Mercury Deposition During the Last 270 Years: A Glacial Ice Core Record of Natural and Anthropogenic Sources." *Environmental Science and Technology* 2002;April 24:2303–2310.

Wikipedia. "*De architectura.*" https://en.wikipedia.org/wiki/De_architectura (accessed August 4, 2016).

Further Reading

Agency for Toxic Substances and Disease Registry (ATSDR), Centers for Disease Control. "Toxicological Profile for Lead." http://www.atsdr.cdc.gov/toxprofiles /tp.asp?id=96&tid=22 (accessed June 4, 2014).

Agency for Toxic Substances and Disease Registry (ATSDR), Centers for Disease Control. "Toxicological Profile for Mercury." http://www.atsdr.cdc.gov/substances /toxsubstance.asp?toxid=24 (accessed June 4, 2014).

Baron, S., M. Lavoie, A. Ploquin, J. Carignan, M. Pulido, and J.-L . De Beaulieu. "Record of Metal Workshops in Peat Deposits: History and Environmental Impact on the Mont Lozere Massif, France." *Environmental Science and Technology* 2005;June 7:1531–5140.

Betts, K. "Dramatically Improved Mercury Removal." *Environmental Science and Technology* 2003;August 1:283–284A.

Burke, M. "Leaded Gasoline Phaseout Becoming a Reality." *Environmental Science and Technology* 2004;September 1:326A.

Cappiello, D. "New Gov't Study Shows Mercury in Fish Widespread." *Huffington Post*, August 19, 2009. http://www.huffingtonpost.com/huff-wires/20090819/us -mercury-contamination (accessed June 4, 2014).

Centers for Disease Control and Prevention (CDC). "Current Trends in Childhood Lead Poisoning—United States: Report to the Congress by the Agency for Toxic Substances and Disease Registry." *Morbidity and Mortality Weekly Report* 1988;37(32):481–485.

Centers for Disease Control and Prevention (CDC). "Lead." http://www.cdc.gov /nceh/lead/ (accessed June 4, 2014).

Centers for Disease Control and Prevention (CDC). "Mercury." http://emergency.cdc .gov/agent/mercury/index.asp (accessed June 4, 2014).

Chemical and Engineering News. "Controlling Mercury Emissions." 2008;May 19:33.

Chemical and Engineering News. "Countries Ink Mercury Pact." 2013;October 21:25.

Chemical and Engineering New. "EPA Hit for Mercury Air Standard." 2005;February 14:24.

Chemical and Engineering News. "EPA's Mercury Rule Challenged." 2005;May 23:30.

Chemical and Engineering News. "FDA, EPA Warn Women on Mercury in Some Fish." 2001;January 22:43.

Chemical and Engineering News. "FDA Prepares New Warning for Mercury in Fish." 2003;December 15:19.

Chemical and Engineering News. "High Levels of Mercury in Lake Fish." 2004; August 9:22.

Chemical and Engineering News. "House Passes Ban on Mercury Exports." 2007; November 19:36.

Chemical and Engineering News. "Lead Cited as Number One Toxic Release." 2005;May 30:27.

Chemical and Engineering News. "Limit on Lead in Air to Be Tightened." 2008; May 12:30.

Chemical and Engineering News. "Mercury Near Toxic Level for Some Women." 2001;March 12:43.

Chemical and Engineering News. "Nearly All Mercury Was Removed from Plant Air Emission, News Summary." 2005;March 7:30.

Chemical and Engineering News. "States Sue to Block EPA Mercury Rule." 2005;April 4:46.

Chemical and Engineering News. "WHO Advisers Urge Reduced Mercury Levels." 2003;July 7:13.

Cooney, C. M. "Water Utilities May Be Stuck with MTBE Cleanup." *Environmental Science and Technology* 2005;July 1:279A.

Crenson, S. L. "Study Records Elevated Mercury." Associated Press, October 20, 2002. http://www.apnewsarchive.com/2002/Study-Records-Elevated-Mercury /id-deba53281a9e89e3b47634d205ece00e (accessed June 4, 2014).

Dalton, L. W. "Methylmercury Toxicology Probed." *Chemical and Engineering News* 2004;January 19:70–71.

Daly, M. "EPA Chief: Climate Plan on Track Despite Mercury Ruling." Phys.org, July 7, 2015. http://phys.org/news/2015-07-epa-chief-climate-track-mercury.html (accessed August 31, 2015).

Endo, T., Y. Hotta, K. Haraguchi, and M. Sakata. "Mercury Concentration in the Red Meat of Whales and Dolphins Marketed for Human Consumption in Japan." *Environmental Science and Technology* 2003;May 7:2681–2685.

EPA. "Landmark U.S. Geological Survey Demonstrates How Methylmercury, Known to Contaminate Seafood, Originates in the Ocean," January 1, 2009. http:// yosemite.epa.gov/opa/admpress.nsf/bd4379a92ceceeac8525735900400c27/577aa da113bdd840852575a900660baf!OpenDocument (accessed June 4, 2014).

EPA. "Lead." http://www2.epa.gov/lead (accessed June 4, 2014).

EPA. "Lead and Copper: A Quick Reference Guide." http://water.epa.gov/lawsregs /rulesregs/sdwa/lcr/upload/LeadandCopperQuickReferenceGuide_2008.pdf (accessed June 3, 2014).

EPA. "Mercury." http://www.epa.gov/mercury (accessed June 4, 2014).

Erickson, B. "Nations Strike Mercury Deal." *Chemical and Engineering News* 2013;January 28:8.

Ethyl Corporation. "Ethyl, the Major Manufacturer of Gasoline Additives, Including Tetraethyl-Lead." http://www.ethyl.com/Pages/default.aspx (accessed June 3, 2014).

Falk, H. "International Environmental Health for the Pediatrician: Case Study of Lead Poisoning." *Pediatrics* 2003;112:259–264.

Girard, J. E. *Principles of Environmental Chemistry,* 2nd ed. (pp. 374–401). Sudbury, MA: Jones and Bartlett, 2010.

Hanson, D. "MTBE: Villain or Victim?" *Chemical and Engineering News* 1999; October18:49.

Hess, G. "EPA Rules Target Mercury Emissions." *Chemical and Engineering News* 2011;March 21:30.

Hess, G. "High Court Weighs EPA Mercury Rule." *Chemical and Engineering News* 2015;May 4:23.

Hileman, B. "CDC Releases Chemicals Survey." *Chemical and Engineering News* 2005;August 1:10.

Hogue, C. "Amount of Mercury in Oceans Rises." *Chemical and Engineering News* 2014;August 11:18.

Hogue, C. "Assessing Mercury: UN Environmental Program Undertakes First Global Study of This Pollutant." *Chemical and Engineering News* 2001;February 19:14.

Hogue, C. "Mercury: Countries Agree to Start Talks to Control the Metal Globally." *Chemical and Engineering News* 2009;March 2:13.

Hogue, C. "Mercury Aired: Lawsuits Are Expected Against EPA Rule to Control Emissions at Power Plants." *Chemical and Engineering News* 2005;March 21:11.

Hogue, C. "Mercury Controls: New Regulation Applied to Chlor-Alkali Producers Using Mercury Cells." *Chemical and Engineering News* 2004;January 5:10.

Hogue, C. "Mercury Option: State and Local Regulators Offer Plan as Alternative to Bush Administration Rule." *Chemical and Engineering News* 2005;November 21:8.

Hogue, C. "Mercury Purge: Congress Is Moving Legislation to Outlaw Mercury-Cell Chlorine Plants." *Chemical and Engineering News* 2009;June 15:24–25.

Hogue, C. "Methanex Loses, U.S. Wins." *Chemical and Engineering News* 2005;September 5:25–28.

Hogue, C. "Schedule Set for Mercury Reductions." *Chemical and Engineering News* 2009;November 2:22.

Japan Times. "Norway Whale Meat Dumped in Japan After Pesticide Finding," March 12, 2015. http://www.japantimes.co.jp/news/2015/03/12/national/norway-whale-meat-dumped-japan-pesticide-finding/#.VeTbPOev13Y (accessed August 31, 2015).

Johnson, J. "Coal-Fired Power Plants: Senate Rejects Bid to Kill Mercury Regulation." *Chemical and Engineering News* 2012;June 25:9.

Johnson, J. "Grappling with Mercury: Little Agreement Found Over Bush Administration Proposal to Cut Mercury from Coal-Fired Utilities." *Chemical and Engineering News* 2004;July 12:19–20.

Johnson, J. "Long Time Cutting: Mercury Emissions Controls May Be Installed by Coal-Fired Utilities After Decades of Delay." *Chemical and Engineering News* 2005;February 28:44–45.

Johnson, J. "Mercury Rising in Pacific Ocean." *Chemical and Engineering News* 2009;May 11:24.

Johnson, J. "Olin to End Hg Releases from Chlor-Alkali Plants." *Chemical Engineering and News* 1999;April 26:8.

Johnson, J. "Too Much of a Bad Thing: As U.S. Companies End Mercury Use, Questions Mount Over Need to Limit World Access to Surplus." *Chemical and Engineering News* 2002;July 29:22–23.

Johnson, J. "Where Goes the Missing Mercury?" *Chemical and Engineering News* 2004;March 15:31–32.

Kauffman, G. B. "Midgley—A Two Time Environmental Loser." *Journal of Chemical Education* 2000;77:1540.

Kovarid, B. "Henry Ford, Charles F. Kettering and the Fuel of the Future." *Automotive History Review* 1998;32:7–27.

Kovarik, W. "Ethyl-Leaded Gasoline: How a Classic Occupational Disease Became an International Public Health Disaster." *International Journal of Occupational Environmental Health* 2005;11:384–397.

LaFraniere, S. "Lead Poisoning in China: The Hidden Scourge." *New York Times,* June 15, 2011. http://www.nytimes.com/2011/06/15/world/asia/15lead.html ?pagewanted=all&_r=0 (accessed 3 June, 2014).

Loeb, A. P. "Birth of the Kettering Doctrine: Fordism, Sloanism and the Discovery of Tetraethyl Lead." *Business and Economic History* 1995;24(1):72–86.

LoGiurato, B. "The Supreme Court Just Handed Obama a Significant Loss on One of His Biggest Environmental Initiatives." *Business Insider,* June 29, 2015. http://www.businessinsider.com/supreme-court-epa-decision-2015-6 (accessed August 31, 2015).

Mason, R. P., M. L. Abbott, R. A. Bodaly, O. R. Bullock Jr., C. T. Driscoll, D. Evers, S. E. Lindberg, M. Murray, and E. B. Swain. "Monitoring the Response to Changing Mercury Deposition." *Environmental Science and Technology* 2005;January 1:15–22A.

Max, A. "Toxins Found in Whales Bode Ill for Humans." *U.S. News,* June 25, 2010. http://www.usnews.com/science/articles/2010/06/25/toxins-found-in-whales-bode -ill-for-humans (accessed June 4, 2014).

McCoy, M. "A Hurried Good Bye to Mercury." *Chemical Engineering and News* 2015;November 2:24–27.

McCoy, M. "PPG Will Remove Lead from Paint." *Chemical Engineering and News* 2016:May 2:15.

New Scientist. "Another Reason to Avoid Mercury." 2006;October14:16.

Organization of Economic Co-operation and Development (OECD) and United Nations Environmental Programme (UNEP). *Phasing Lead Out of Gasoline: An Examination of Policy Approaches in Different Countries.* Paris: Author, 1999. http://walshcarlines.com/pdf/unepgas.pdf (accessed June 4, 2014).

Pacyna, E. G., and J. M. Pacyna. "Global Emission of Mercury from Anthropogenic Sources in 1995." *Water, Air, and Soil Pollution* 2002;137:149–165.

Reddy, A., and C. L. Braun. "Lead and the Romans." *Journal of Chemical Education* 2010;87(10):1052–1055.

Reisch, M. S. "Bromine Comes to the Rescue Mercury Power Plants Emissions." *Chemical and Engineering News* 2015;March 16:17–19.

Reisch, M. S. "Getting Rid of Mercury." *Chemical and Engineering News* 2008; November 24:22–23.

Reisch, M. S. "Getting the Lead Out." *Chemical and Engineering News* 2006;April 24:26.

Renner, R. "Mercury Woes Appear to Grow." *Environmental Science and Technology* 2004;April 15:144A.

Renner, R. "Rethinking Atmospheric Mercury." *Environmental Science and Technology* 2004;December 1:448–449A.

Rosner, D., and G. Markowitz. "A 'Gift of God'?: The Public Health Controversy Over Leaded Gasoline During the 1920s." *American Journal of Public Health* 1985;75(4):344–352.

Rouhi, A. M. "Mercury Showers." *Chemical and Engineering News* 2002;April 15:40.

Ryan, J. A., K. G. Scheckel, W. R. Berti, S. L. Brown, S. W. Castell, R. L. Chaney, J. Hallfrisch, M. Doolan, P. Grevantt, M. Maddaloni, and D. Mosby. "Reducing Children's Risk from Lead in Soil." *Environmental Science and Technology* 2004;January 1:19–24A.

Saar, R. A. "New Efforts to Uncover the Dangers of Mercury. *New York Times,* November 2, 1999, p. D7.

Sanburn, J. "Lead in Flint's Water Is Finally Below Federal Limit, Study Finds." *Time,* 2016:August 11. http://time.com/4448573/flint-water-lead-levels-improve-study (accessed August 16, 2016).

Sanburn, J. "The Toxic Tap." *Time* 2016:187(3):32–39.

Schnoor, J. L. "Comment: The Case Against Mercury." *Environmental Science and Technology* 2004;February 1:47A.

Seigneur, S., D. Vijayaraghavan, K. Lohman, P. Karamchandani, and C. Scott. "Global Source Attribution for Mercury Deposition in the United States." *Environmental Science and Technology* 2004;December 3:555–569.

Stockton, N. "How the EPA Puts a Price Tag on Pollution." *Wired,* January 1, 2015. http://www.wired.com/2015/07/epa-puts-price-tag-pollution (accessed August 31, 2015).

Tremblay, J. F. "Weaning China Off Mercury." *Chemical and Engineering News* 2016;January 18:22–23.

Twealt, S. J., L. J. Bragg, and R. B. Finkelman. "Mercury in U.S. Coal—Abundance, Distribution and Modes of Occurrence." Washington, DC: U.S. Geological Survey, September 2001, USGS Fact Sheet FS-095-01.

U.S. Geological Survey. "Mercury Contamination of Aquatic Ecosystems." http://water.usgs.gov/wid/FS_216-95/FS_216-95.html (accessed June 4, 2014).

Warren, C. *Brush with Death: A Social History of Lead Poisoning.* Baltimore: Johns Hopkins University Press, 2000.

Wikipedia. "List of Countries by Coal Production." http://en.wikipedia.org/wiki /List_of_countries_by_coal_production (accessed September 26, 2014).

Wikipedia. "Metal Toxicity." http://en.wikipedia.org/wiki/Metal_toxicity (accessed June 4, 2014).

World Coal Association. "Resources." http://www.worldcoal.org/resources/ (accessed September 26, 2014).

World Health Organization. "Lead." http://www.who.int/ipcs/assessment/public _health/lead/en.

World Health Organization. "Mercury." http://www.who.int/mediacentre/factsheets /fs361/en (accessed June 4, 2014).

Wright, K. "Our Preferred Poison." *Discover* 2005;March:58–65.

4. Elimination of Chlorinated Hydrocarbons from Our Environment

Notes

Carson, R. *Silent Spring.* Boston: Houghton Mifflin, 1962.

Dirtu, A., V. L. B. Jaspers, R. Cernat, H. Neels, and A. Covaci. "Distribution of PCBs, Their Hydroxylated Metabolites, and Other Phenolic Contaminants in Human Serum from Two European Countries." *Environmental Science and Technology* 2010;October 13:2876–2883.

Hogberg, J., A. Hanberg, M. Berglund, S. Skerfving, M. Remberger, A. M. Calafat, A. F. Filipsson, B. Jansson, N. Johansson, M. Appelgren, and H. Hakansson. "Phthalate Diesters and Their Metabolites in Human Breast Milk, Blood or Serum, and Urine as Biomarkers of Exposure in Vulnerable Populations." *Environmental Health Perspectives* 2008;116(3):334–339.

Jacobs, J. P. "Sippy Cups, Baby Bottles, Now BPA-free—Industry." E & E Publishing LLC, October 7, 2011. http://www.eenews.net/stories/1059954726 (accessed August 4, 2016).

Kinkela, D. *DDT and the American Century: Global Health, Environmental Politics, and the Pesticide That Changed the World.* Chapel Hill: University of North Carolina Press, 2011.

Mendiola, J., N. Jorgensen, A. M. Andersson, A. M. Calafat, S. Ye, J. B. Redmon, E. Z. Drobnis, C. Want, A. Sparks, S. W. Thurston, F. Liu, and S. Swan. "Are Environmental Levels of Bisphenol A Associated with Reproduction Function in Fertile Men?" *Environmental Health Perspectives* 2010;118(9):1286–1291.

Pagana, K. D., and T. J. Pagana. *Manual of Diagnostic and Laboratory Tests.* St. Louis: Mosby, 1998.

"Paracelsus." http://www.brainyquote.com/quotes/authors/p/paracelsus.html (accessed August 4, 2016).

Takeuchi, T., and O. Tsutsumi. "Serum Bisphenol A Concentrations Showed Gender Differences, Possibly Linked to Androgen Levels." *Biochemical and Biophysical Research Communications* 2002;291(1):76–78.

Woodwell, G. M., C. F. Wurster, and P. A. Isaacson. "DDT Residues in an East Coast Estuary: A Case of Biological Concentration of a Persistent Insecticide." *Science* 1967;156:821–824.

You, L., X. Zhu, M. J. Shrubsole, H. Fan, J. Chen, J. Dong, C. M. Hao, and Q. Dal. "Renal Function, Bisphenol A, and Alkylphenols: Results from the National Health and Nutrition Examination Survey (NHANES 2003–2006)." *Environmental Health Perspectives* 2011;119(4):527–533.

Further Reading

Alonso-Zaldivar, R., and L. Tanner. "FDA Defends Plastic Linked with Health Risks." *USA Today*, September 16, 2008. http://usatoday30.usatoday.com/news/health/2008-09-16-2758258431_x.htm (accessed July 28, 2014).

Bailey, R. A., Herbert M. Clark, J. P. Ferris, S. Krause, and R. L. Strong. *Chemistry of the Environment* (pp. 223–291). New York: Academic, 2002.

Baird, C. *Environmental Chemistry* (pp. 193–284). New York: Freeman, 1995.

Baum, R. M. "Meet Joe Chemical." *Chemical Engineering and News* 2012;June 25:3.

Betts, K. "Flame Retardant Travels the Globe." *Chemical Engineering and News* 2010;November 15:8.

Betts, K. S. "Improving Fish Food." *Environmental Science and Technology* 2004;March 1:88A.

Blahut, D. "Is That Plastic Container Safe?" *Woman's Day*, September 20, 2010. http://www.womansday.com/health-fitness/is-that-plastic-container-safe-110601 (accessed July 29, 2014).

Bomgardner, M. M. "No Easy Fix for Food Can Coatings." *Chemical Engineering and News* 2013;February 11:24–25.

Braga, O., G. A. Smythe, A. I. Schafer, and A. J. Feitz. "Fate of Steroid Estrogens in Australian Inland and Coastal Wastewater Treatment Plants." *Environmental Science and Technology* 2005;March 18:3351–3358.

Burke, M. "POPs Treaty Takes Flight." *Environmental Science and Technology* 2004;May 1:157A.

Chemical Engineering and News. "Five Chemicals Pass Hurdle for Control Under POPs Treaty." 2005;November 28:23.

Chemical Engineering and News. "Persistent Pollutant Pact Enters Into Force." 2004;May 24:24.

Chemical Engineering and News. "San Francisco Bans Phthalates, Bisphenol A." 2006;June 12:28.

Christen, K. "U.N. Negotiations on POPs Snag on Malaria." *Environmental Science and Technology* 1999;November 1:444–445A.

Dalton, L. "Salmon Move PCBs." *Chemical Engineering and News* 2003;September 22:10.

Du Yeon, Bang, Minji Kyung, Min Ji Kim, Bu Young Jung, Myung Chan Cho, Seul Min Choi, et al. "Human Risk Assessment of Endocrine-Disrupting Chemicals Derived from Plastic Food Containers." *Comprehensive Reviews in Food Science and Food Safety* 2012;11(5):453–470.

Dunham, W. "Plastic Bottle Chemical May Be Harmful: Agency." Reuters, April 15, 2008. http://www.reuters.com/article/2008/04/15/us-plastic-bottles-idUSN15139 29320080415 (accessed July 28, 2014).

EPA. "Endocrine Disruption." http://www.epa.gov/endo (accessed July 29, 2014).

Erickson, B. E. "Bisphenol A Battle." *Chemical Engineering and News* 2008;November 17:42–45.

Erickson, B. E. "Endocrine Disrupter Found in Aircraft Deicer." *Environmental Science and Technology* 2003;October 1:345–346A.

Erickson, B. E. "EPA Expands Endocrine Disruptor Screen Program." *Chemical Engineering and News* 2010;November 22:24.

Erickson, B. E. "EPA Retools Endocrine Program." *Chemical Engineering and News* 2013;March 18:30.

Erickson, B. E. "EPA Targets Bisphenol A." *Chemical Engineering and News* 2010;April 5:8.

Erickson, B. E. "FDA Acts on Bisphenol A Petitions." *Chemical Engineering and News* 2012;June 18:28.

Erickson, B. E. "FDA Bans BPA in Baby Bottles." *Chemical Engineering and News* 2012;July 23:23.

Erickson, B. E. "Levels of Bisphenol A in U.S. Population Drops." *Chemical Engineering and News* 2012;October 8:30.

Erickson, B. E. "Pressure on Plasticizers." *Chemical Engineering and News* 2015;June 22:11–15.

Everts, S. "Chemicals Leach from Packaging." *Chemical Engineering and News* 2009;August 31:11–15.

Everts, S. "Drugs in the Environment." *Chemical Engineering and News* 2010;March 29:23–24.

Farkas, B. "Health Screening for Teflon to Start." *Free Republic,* July 9, 2005. http://www.freerepublic.com/focus/f-news/1439535/posts (accessed July 28, 2014).

Favole, J. A. "FDA to Revisit Decision on Safety of BPA." *Wall Street Journal,* June 3, 2009. http://www.wsj.com/articles/SB124405286248681991 (accessed December 16, 2014).

Fox, M. "Hormone Experts Worried About Plastics, Chemicals." Reuters, June 10, 2009. http://www.reuters.com/article/us-bisphenol-idUSTRE55A0JK20090611 (accessed July 29, 2014).

Girard, J. E. *Principles of Environmental Chemistry* (pp. 402–461). Sudbury, MA: Jones and Bartlett, 2010.

Gorman, J. "A Drug Used for Cattle Is Said to Be Killing Vultures." *New York Times International*, January 29, 2004.

Guenter, K., V. Heinke, B. Thiele, E. Kleist, H. Prast, and T. Raecker. "Endocrine Disrupting Nonylphenols Are Ubiquitous in Food." *Environmental Science and Technology* 2002;March 19:1676–1680.

Halford, B. "Side Effects." *Chemical Engineering and News* 2008;February 25:13–17.

Hanson, D. J. "Bisphenol A Called Mostly Safe." *Chemical Engineering and News* 2008;April 21:11.

Hileman, B. "Bisphenol A Harms Mouse Eggs." *Chemical Engineering and News* 2003;April 7:7.

Hileman, B. "California Bans Phthalates in Toys." *Chemical Engineering and News* 2007;October 22:12.

Hileman, B. "Clash of Views on Bisphenol A." *Chemical Engineering and News* 2003;May 5:40–41.

Hileman, B. "DEHP Found in Baby Products." *Chemical Engineering and News* 2005;November 14:34.

Hileman, B. "EU Bans Three Phthalates from Toys, Restricts Three More." *Chemical Engineering and News* 2005;July 11:11.

Hileman, B. "Malaria Control." *Chemical Engineering and News* 2006;July 24:30–31.

Hileman, B. "Phthalates and Male Babies." *Chemical Engineering and News* 2005;June 6:8.

Hileman, B. "Tracking the Chemicals in Us." *Chemical Engineering and News* 2007;April 23:32–34.

Hogue, C. "Bisphenol A Restrictions." *Chemical Engineering and News* 2008;October 27:8.

Hogue, C. "Estimating Properties of New Chemicals: EPA Software Undergoes Peer Review." *Chemical Engineering and News* 2006;April 3:32–33.

Hogue, C. "PFOA Called Likely Human Carcinogen." *Chemical Engineering and News* 2005;July 4:5.

Hogue, C. "Senate Panel Adopts Chemicals Reform Bill." *Chemical Engineering and News* 2012;July 30:9.

Hogue, C. "U.S. Vote at Treaty Meetings Threatened." *Chemical Engineering and News* 2004;March 29:22–23.

Houck, O. "Tales from a Troubled Marriage: Science and Law in Environmental Policy." *Science* 2003;302:1926–1929.

Jacobs, M. "Chemistry and Democracy." *Chemical Engineering and News* 1999;September 20:5.

Koplin, D., E. T. Furlong, M. T. Meyer, E. M. Thurman, S. D. Zaugg, L. B. Barber, and H. T. Buxton. "Pharmaceuticals, Hormones, and Other Organic Wastewater Contaminants in U.S. Streams, 1999–2000: A National Reconnaissance." *Environmental Science and Technology* 2002;March 13:1202–1211.

Mayo Clinic. "Natural Endocrine Concentrations." http://www.mayomedicallabora tories.com/test-catalog/alphabetical/P (accessed March 18, 2015).

McCoy, M. "The End of an Era." *Chemical Engineering and News* 2009;July 20:30.

Mellanby, K. *The DDT Story*. Hampshire, UK: The British Crop Protection Council, 1992.

Morrison, J. "Water Pollution: EPA Would Ban Hospitals from Dumping Some Unused Drugs Down the Drain." *Chemical Engineering and News* 2015;September 7:11.

Mullin, R. and S. Morrissey "Momentum Builds Against Bisphenol A." *Chemical Engineering and News* 2008;April 28:11.

National Health and Nutrition Examination Survey (NHANES II). http://www.cdc .gov/nchs/nhanes/nhanes1999–2000/questionnaires99_00.htm (accessed December 16, 2014).

National Institutes of Health; National Institute of Environmental Health Science. 1999–2000 Survey, http://www.niehs.nih.gov/health/topics/agents/endocrine; 2009–2010 survey, http://wwwn.cdc.gov/nchs/nhanes/search/nhanes09_10.aspx (accessed July 29, 2014).

National Resources Defense Council. "Endocrine Disruptors." http://www.nrdc .org/health/effects/qendoc.asp (accessed July 29, 2014).

New Scientist. "Boys Beware of Moisturizer." 2007;February 3:7.

New Scientist. "Can India's Vultures Be Saved?" 2006;June 3:5.

New Scientist. "Drastic Plastic Ban." *New Scientist* 2005;April 23:6.

New Scientist. "Drug Firm Backs Vultures." 2006;August 12:6.

North American Commission for Environmental Cooperation. "History of DDT in North America to 1997." http://www3.cec.org/islandora/en/item/1620-history -ddt-in-north-america-1997-and-1996-presentation-mexican-ministry-en.pdf (accessed June 10, 2014).

Renner, R. "Controversy Clouds Atrazine Studies." *Environmental Science and Technology* 2004;March 15:107–108A.

Rettner, R. "Snow at Highest Elevations No Longer Pure." *Live Science*, December 10, 2009. http://www.livescience.com/10602-snow-highest-elevations-longer-pure .html (accessed June 10, 2014).

Ritter, S. K. "Bisphenol A." *Chemical Engineering and News* 2011;June 6:11–19.

Ritter, S. K. "Designing Away Endocrine Disruption." *Chemical Engineering and News* 2012;December 17:33.

Ritter, S. K. "Measuring Persistence." *Chemical Engineering and News* 2015; March 2:10–13.

Ritter, S. K. "Red Flag Raised for BPA Metabolite." *Chemical Engineering and News* 2012;October 8:43.

Ritter, S. K. "Xenoestrogens Taint Food Additives." *Chemical Engineering and News* 2009;January 19:57.

Rowan, K. "Soaring BPA Levels Found in People Who Eat Canned Foods." Fox News, November 23, 2011. http://www.foxnews.com/health/2011/11/23/soaring -bpa-levels-found-in-people-who-eat-canned-foods (accessed July 28, 2014).

Rudel, R. A., D. E. Camann, J. D. Spengler, L. R. Korn, and J. G. Brody. "Phthalates, Alkylphenols, Pesticides, Polybrominated Biphenyl Esters, and Other Endocrine-Disrupting Compounds in Indoor Air and Dust." *Environmental Science and Technology* 2003;September 13:4543–4553.

Sax, L. "Polyethylene Terephthalate May Yield Endocrine Disruptors." *Environmental Health Perspective* 2010;118(4):445–448.

Short, P. L. "European Gardeners' Shrinking Arsenal." *Chemical Engineering and News* 2002;December 9:18–19.

Slater, D. "Chemical Reactions: Here's an Idea, Let's Not Test Toxics on the Kids." *Sierra*. 2010;September/October:18.

Stokstad, E. "Pollution Gets Personal." *Science* 2004;304:1892–1894.

Tepper, R. "Another Reason to Stop Eating Processed Foods." *Yahoo! News*, February 20, 2014. https://www.yahoo.com/food/another-reason-to-stop-eating -processed-foods-77295344197.html (accessed July 28, 2014).

Trivedi, B. "Hormones in the Water Devastate Wild Fish." *New Scientist* 2007; May 26:16.

Tullo, A. H. "Babies on Board." *Chemical Engineering and News* 2009;August 31:20.

Tullo, A. H. "Food Firms Push to Phase Out Bisphenol A." *Chemical and Engineering News* 2016:April 4:11.

Vanderberg, L. N., R. Hauser, M. Marcus, N. Olea, and W. V. Welshons. "Human Exposure to Bisphenol A (BPA)." *Reproductive Toxicology*, 2007;24:139–177.

Voith, M. "Removing Bisphenol A." *Chemical Engineering and News* 2010;October 25:10.

Webster, P. "Study Finds Heavy Contamination Across Vast Russian Arctic." *Science* 2004;306:1875.

5. The Safety of Chemicals in Our Food and Water

Notes

Borenstein, S. "An American Life Worth Less Today." *USA Today*, July 11, 2008. http://usatoday30.usatoday.com/news/nation/2008-07-10-796349025_x.htm.

Gadgil, A. "Drinking Water in Developing Countries." *Annual Review of Energy and the Environment* 2006;23:253–286.

"Paracelsus." http://www.brainyquote.com/quotes/authors/p/paracelsus.html (accessed August 4, 2016).

Wilson, R., and A. C. Crouch. "Risk Assessment and Comparisons: An Introduction." *Science* 1987;263:267–270.

Further Reading

American Cancer Society. *Cancer Facts and Figures 2013*. Atlanta: American Cancer Society, 2013.

Arnaud, C. H. "The Exposome Turns 10." *Chemical Engineering and News* 2015;September 28:41.

Baird, C. *Environmental Chemistry* (pp. 215–284). New York: Freeman, 1995.

Calabrese, E. J. "Animal Extrapolation: A Look Inside the Toxicologist's Black Box." *Environmental Science and Technology* 1987;July 1:618–623.

Centers for Disease Control and Prevention (CDC). "Health, United States, 2013— At a Glance." http://www.cdc.gov/nchs/hus/at_a_glance.htm (accessed July 29, 2014).

Centers for Disease Control and Prevention (CDC). "National Report on Human Exposure to Environmental Chemicals." https://www.cdc.gov/exposurereport (accessed August 11, 2016).

Chiuchiolo, A. L., R. M. Dickhut, M. A. Cochran, and H. W. Ducklow. "Persistent Organic Pollutants at the Base of the Antarctic Marine Food Web." *Environmental Science and Technology* 2004;May 29:3551–3557.

EPA. *Fact Sheet: Contaminated Sediment: EPA's Report to Congress.* Washington, DC: Environmental Protection Agency, 1998. EPA-823-F-98-001.

EPA. "Human Health Risk Assessment." http://www.epa.gov/risk_assessment/health -risk.htm (accessed July 29, 2014).

Erickson, B. E. "Endocrine Disrupter Found in Aircraft Deicer." *Environmental Science and Technology* 2003;October 1:345–346A.

Erickson, B. E. "EPA Releases Endocrine Screening List." *Chemical Engineering and News* 2013;June 24:28.

Erickson, B. E. "EPA Retools Endocrine Program." *Chemical Engineering and News* 2013;March 18:30–32.

Erickson, B. E. "Next-Generation Risk Assessment." *Chemical and Engineering News* 2009;June 22:30–33.

Fowle, J. R. III. "Improving Toxicity Testing for Better Decision-Making." *Chemical Engineering and News* 2015;July 13:31.

Girard, J. E. *Principles of Environmental Chemistry*, 2nd ed. (pp. 404–439). Sudbury, MA: Jones and Bartlett, 2010.

Gough, M. "Estimating Cancer Mortality." *Environmental Science and Technology* 1989;August 1:925–930.

Hileman, B. "Phthalates and Male Babies." *Chemical Engineering and News* 2005;June 6:8.

Hileman, B. "Tracking the Chemicals in Us." *Chemical Engineering and News* 2007;April 23:32–34.

Hogue, C. "Estimating Properties of New Chemicals: EPA Software Undergoes Peer Review." *Chemical Engineering and News* 2006;April 3:32–33.

Hogue, C. "PFOA Called Likely Human Carcinogen." *Chemical Engineering and News* 2005;July 4:5.

Hogue, C. "Senate Panel Adopts Chemicals Reform Bill." *Chemical Engineering and News* 2012;July 30:9.

Krimsky, S., and A. Plough. *Social Battles on Environmental Risks.* Dover, MA: Auburn House, 1989.

Louvar, J. F., and B. D. Louvar. *Health and Environmental Risk Assessment: Fundamentals with Applications* (pp. 236–254). Upper Saddle River, NJ: Prentice-Hall, 1998.

Michaels, D., and C. Montforton. "Manufacturing Uncertainty: Contested Science and the Protection of the Public's Health and Environment." *Public Health Matters* 2005;95(S1):S39–45.

New Scientist. "Boys Beware of Moisturizer." 2007;February 3:7.

Ritter, S. K. "Designing Away Endocrine Disruption." *Chemical Engineering and News* 2012;December 17:33.

Ropeik, D. *How Risky Is It, Really?* New York: McGraw-Hill, 2010.

Rudel, R. A., D. E. Camann, J. D. Spengler, L. R. Korn, and J. G. Brody. "Phthalates, Alkylphenols, Pesticides, Polybrominated Biphenyl Esters, and Other Endocrine-Disrupting Compound in Indoor Air and Dust." *Environmental Science and Technology* 2003;September 13:4543–4553.

Stevens, J. B., and D. L. Swackhamer. "Environmental Pollutions: A Multimedia Approach to Modeling Human Exposure." *Environmental Science and Technology* 1989;October 1:1180–1186.

Suter, G. W. "Ecological Risk Assessment in the United States Environmental Protection Agency: A Historical Overview." *Integrated Environmental Assessment and Management* 2008;4(3):285–289.

Van Straalen, N. M. "Ecotoxicology Becomes Stress Ecology." *Environmental Science and Technology* 2003;September 1:326–330A.

Wilkinson, C. F. "Being More Realistic About Chemical Carcinogenesis." *Environmental Science and Technology* 1987;21(9):843–847.

6. Saving Our Atmosphere for Our Children

Notes

Farman, J. C., B. G. Gardiner, and J. D. Shanklin. "Large Losses of Total Ozone in Antarctica Reveal Seasonal CLOx/NOx Interaction." *Nature* 1985;315:207–210.

Lewis, Drew. http://izquotes.com/quote/111650 (accessed November 3, 2016).

Masters, J. "Weather Underground: The Skeptics Versus the Ozone Hole." https://www.wunderground.com/resources/climate/ozone_skeptics.asp (accessed August 8, 2016).

Molina, M. J., and Rowland, F. S. "Stratospheric Sink for Chlorofluoro-
methanes: Chlorine Atom Catalyzed Destruction of Ozone." *Nature* 1974;249:
810–814.

Sagan, C. "United Nations Environmental Programme, June 22, 2016." http://www
.unep.org/NewsCentre/default.aspx?DocumentID=27080&ArticleID=36235
(accessed August 8, 2016).

Steinbeck, J. "Proverbia." http://en.proverbia.net/citasautor.asp?autor=16942 (accessed
August 8, 2016).

Wald, M. L. "No Headline." *New York Times,* September 14, 1985. http://www.nytimes
.com/1985/09/14/us/no-headline-003213.html (accessed August 8, 2016).

Further Reading

Acid Rain Program in Practice. http://www.colorado.edu/economics/morey/8545
/student/so2permits/practice.html (accessed July 30, 2014).

American Chemical Society. *Chemistry in Context: Applying Chemistry to Society,*
6th ed. (pp. 56–99). New York: McGraw-Hill, 2009.

Baird, C. *Environmental Chemistry* (pp. 90–99). New York: Freeman, 1995.

Benedick, R. E. *Ozone Diplomacy: New Directions in Safeguarding the Planet.* Cambridge, MA: Harvard University Press, 1998.

Boyd, J., D. Burtraw, A. Krupnick, V. McConnell, R. G. Newell, K. Palmer, J. N.
Sanchirico, and M. Walls. "Trading Cases." *Environmental Science and Technology* 2003;June 1:216–222A.

Briney, A. "Acid Rain: The Causes, History, and Effect of Acid Rain." About.com.
http://geography.about.com/od/globalproblemsandissues/a/acidrain.htm (accessed July 30, 2014).

Brown, P. "Ozone Layer Most Fragile on Record." *Guardian,* April 27, 2005. http://
www.theguardian.com/science/2005/apr/27/environment.research (accessed July
30, 2014).

Burtraw, D., and E. Mansur. "Environmental Effects of SO_2 Trading and Banking."
Environmental Science and Technology 1999;August 31:3489–3494.

Carlozicz, M. "The Ozone Layer: Our Global Sunscreen." *ChemMatters.* April 2013.
www.acs.org/chemmatters.

Christen, K. "Growing Ozone Hole 'Hangover.'" *Environmental Science and Technology* 2005;June 15:255A.

DeMatto, A. "The Downside to the Recovery of the Ozone Hole." *Live Science,* July 1,
2010. http://www.livescience.com/6671-downside-recovery-ozone-hole.html
(accessed August 7, 2014).

Dismukes, G. C., V. V. Klimov, S. V. Baranov, Y. N. Kozlov, J. DasGupta, and A.
Tyryshkin. "The Origin of Atmospheric Oxygen on Earth: The Innovation of Ox-
ygenic Photosynthesis." *Proceedings of the National Academy of Sciences*
2001;98(5):2170–2175.

Environmental Defense Fund. "The Power of Markets to Help the Planet." http://www.edf.org/approach/markets/acid-rain (accessed July 30, 2014).

EPA. "Acid Rain in New England." http://www.epa.gov/region1/eco/acidrain (accessed July 30, 2014).

Girard, J. E. *Principles of Environmental Chemistry* (pp. 198–203). Sudbury, MA: Jones and Bartlett, 2010.

Hecht, J. "Snug as a Bug: Bacteria Enveloped the Ancient Earth in a Warming Blanket of Gas." *New Scientist* 2000;June 17:18–19.

Heilprin, J. "Ozone Layer Faces Record Loss Over Arctic." *USA Today*, April 5, 2011. http://usatoday30.usatoday.com/news/world/environment/2011-04-05-ozone-arctic_N.htm (accessed July 30, 2014).

Hess, G. "EPA Moves Toward Ban on Certain Hydrofluorocarbons." *Chemical Engineering and News* 2014;July 31:5.

Hileman, B. "Protecting Ozone Helps Protect Climate." *Chemical Engineering and News* 2005;May 2:28–29.

Hogue, C. "Clampdown on HFCs." *Chemical Engineering and News* 2015; July 13:4.

Hogue, C. "U.S., China Agree on Refrigerants." *Chemical Engineering and News* 2013;June 17:8.

Holland, H. D. "Evidence of Life on Earth More than 3850 Million Years Ago." *Science* 1997;275:38 39.

Houser, G. "Cap and Trade: A Fatal Distraction." CommonDreams.org, November 30, 2009. http://www.commondreams.org/views/2009/11/30/cap-and-trade-fatal-distraction (accessed July 30, 2014).

Kandel, R. *Water from Heaven: The Story of Water from the Big Bang to the Rise of Civilization and Beyond* (pp. 11–28). New York: Columbia University Press, 1998.

Laws, E. A. *Aquatic Pollution: An Introductory Text* (pp. 539–563). 3rd ed. New York: Wiley, 2000.

McFarland, M. "Chlorofluorocarbons and Ozone." *Environmental Science and Technology* 1989;23(10):1203–1207.

Morrisette, P. M. "The Evolution of Policy Responses to Stratospheric Ozone Depletion." *Natural Resources Journal* 1989;(29):793–820.

Newton, D. E. *The Ozone Dilemma*. Contemporary World Issues series. Santa Barbara, CA: ABC-CLIO, 1995.

O'Sullivan, D., and P. S. Zurer "Saving the Ozone Layer: Key Issues Face Tough Negotiations." *Chemical Engineering and News* 1989;March 13:4–5.

O'Sullivan, D. A. "International Gathering Plans Ways to Safeguard Atmospheric Ozone." *Chemical Engineering and News* 1989;June 26:33–36.

Reish, M. "CFC Producers Race to Bring on Substitutes." *Chemical Engineering and News* 1989;July 3:8.

Reish, M., and P. S. Zurer. "CFC Production: Du Pont Seeks Total Phaseout." *Chemical Engineering and News* 1988;April 4:4–5.

Ritter, S. K. "Earth's Ozone Hole Is Healing." *Chemical Engineering and News* 2015;June 1:8.

Tullo, A. H. "The Switch Is On for Refrigerants." *Chemical Engineering and News* 2006;April 24:24–25.

U.S. Exchange Rate and Commodity Process. "Reducing Acid Rain Emissions in the US Through the Sulfur Tax." http://blogs.ubc.ca/lisatam/2013/03/08/reducing -acid-rain-emissions-in-the-us-through-the-sulfur-tax (accessed July 30, 2014).

Voice of America (VOA). "Ozone Layer on Road to Recovery," September 10, 2014. http://www.voanews.com/content/ozone-layer-on-road-to-recovery/2445329 .html (accessed August 31, 2015).

Wikipedia. "Atmosphere of Earth." http://en.wikipedia.org/wiki/Atmosphere_of _Earth (accessed July 30, 2014).

Zurer, P. S. "Arctic Ozone Loss: Fact-Finding Mission Concludes Outlook Is Bleak." *Chemical Engineering and News,* 1989;March 6:29–33.

Zurer, P. S. "As CFC Ban Quietly Comes Into Force, Attention Turns to Other Concerns." *Chemical Engineering and News* 1995;December 4:26–27.

Zurer, P. S. "CFC Phaseout Feasible by 2000, Panel Says." *Chemical Engineering and News* 1989;August 14:5.

Zurer, P. S. "CFC Production Cuts: EPA Rules Already Under Attack." *Chemical Engineering and News* 1988;August 8:3.

Zurer, P. S. "CFC Substitutes: Candidates Pass Early Toxicity Tests." *Chemical Engineering and News* 1989;October 9:4.

Zurer, P. S. "International Effort to Examine Arctic Ozone Loss Gets Under Way." *Chemical Engineering and News* 1989;January 2:30.

Zurer, P. S. "Meeting to Explore Faster CFC Phaseout." *Chemical Engineering and News* 1989;April 24:5–6.

Zurer, P. S. "Ozone Layer: Study Finds Alarming Global Losses." *Chemical Engineering and News* 1988;March 21:6–7.

Zurer, P. S. "Producers, Users Grapple with Realities of CFC Phaseout." *Chemical Engineering and News* 1989;July 24:7–13.

Zurer, P. S. "Sacrificing CFCs to Save the Ozone Layer." *Chemical Engineering and News* 1988;November 14:59–60.

Zurer, P. S. "Search Intensifies for Alternatives to Ozone-Depleting Halocarbons." *Chemical Engineering and News* 1988;February 8:17–20.

Zurer, P. S. "Stratospheric Zone Shows Global Decrease." *Chemical Engineering and News* 1988;January 11:21.

Zurer, P. S. "Studies on Ozone Destruction Expand Beyond Antarctic." *Chemical Engineering and News* 1988;May 30:16–25.

7. Legislating Industry

Notes

Baum, R. M. "Scientists' Discontent with Bush." *Chemical and Engineering News,* 2006;May 8:5.

Beecher, H. W. http://www.allgreatquotes.net/1/a-law-is-valuable-not-because-it-is -law-but-because-there-is-right-in-it (accessed August 9, 2016).

Borenstein, S. "AP Analysis: VW Evasion Likely Led to Dozens of Deaths." *Yahoo! Finance,* October 5, 2015. http://finance.yahoo.com/news/ap-analysis-vw -evasion-likely-led-dozens-deaths-074550966--finance.html (accessed November 24, 2015).

BushWatch. http://www.bushwatch.org (accessed September 26, 2014).

Carson, R. *Silent Spring.* Boston: Houghton Mifflin, 1962.

Chemical and Engineering News. "ACLU Blames Bush for Curbing Scientific Freedoms." 2005;June 27:33.

Chemical and Engineering News. "Members of EPA Chemicals Panel Resign." 2006a;October 9:26.

Chemical and Engineering News. "NASA Criticized for Censoring Scientist." 2006d;February 6:19.

Chemical and Engineering News. "Science Advisers at NASA Resign." 2006c;August 28:24.

Chemical and Engineering News. "Science Advisers Rebuke EPA on Particulate Rule." *Chemical and Engineering News* 2006b;October 9:26.

CNN. "EPA E-Mail to Workers: Don't Answer Inspector's Questions," July 28, 2008. http://www.cnn.com/2008/POLITICS/07/28/epa.gag.order/index.html?iref =mpstoryview (accessed August 21, 2014).

Donne, J. "No Man Is an Island." http://www.poemhunter.com/poem/no-man-is-an -island (accessed August 9, 2016).

Environmental Science and Technology. "Politically Distorted Science." 2003;November 1:377A.

Erickson, B. "President Signs TSCA Reform Bill Into Law." *Chemical and Engineering News* 2016;June 27:15.

Erlich, P. *The Population Bomb.* Cutchogue, NY: Buccaneer Books, 1995.

Hebert, H. J. "EPA Scientists Complain About Political Pressure." *USA Today,* April 23, 2008. http://usatoday30.usatoday.com/news/washington/2008-04-23- 1846012511_x.htm (accessed August 21, 2014).

Hess, G. "Bodman Abolishes DOE Advisory Board." *Chemical and Engineering News* 2006;April 17:8.

Hogue, C. "Changing the Rules on Regulations." *Chemical and Engineering News* 2007;January 29:10.

Hogue, C. "Climate Debate Turns Foul: Scientists Are Being Intimidated and Harassed Because of Their Research, They Tell Congress." *Chemical and Engineering News* 2010;June 7:31.

Hogue, C. "Data Snarl." *Chemical and Engineering News,* 2012:May 28:44–45.

Hogue, C. "EPA Analysis Called Faulty." *Chemical and Engineering News* 2005;December 12:8.

Hogue, C. "Smog, Policy, and Chemistry." *Chemical and Engineering News* 2005;January 24:25–26.

Hogue, C. "Support Grows for Chemical Law Reform." *Chemical and Engineering News* 2013;June 10:22–23.

National Resources Defense Council. "Rewriting the Rules: The Bush Administration's First Term Environmental Record," January 19, 2005. http://www.nrdc.org/bushrecord (accessed September 26, 2014).

New Scientist. "Carbon Omissions." 2007;March 10:5.

New Scientist. "Chemical Progress." 2006;December 6:7.

Pearce, F. "Climate Report 'Was Watered Down.'" *New Scientist* 2007;March 10:10.

Petkewich, R. "House Panel Sees Gap in Technical Advice." *Chemical and Engineering News* 2006;August 14:46–48.

Randerson, J. "Should Governments Play Politics with Science?" *New Scientist* 2004;October 9:12–14.

Renner, R. "White House Denies Scientific Manipulation" *Environmental Science and Technology* 2004;April 15:142A.

Revkin, A. "Ex-Bush Aide Who Edited Climate Reports to Join ExxonMobil." *New York Times* 2005;June 15.

SierraClub.org. "The Bush Record: More than 300 Crimes Against Nature," January/February 2006. http://vault.sierraclub.org/sierra/200409/bush_record_print.asp (accessed September 26, 2014).

UN Framework Convention on Climate Change. "Status of Ratification of the Kyoto Protocol." https://unfccc.int/kyoto_protocol/status_of_ratification/items/2613.php (accessed September 26, 2014).

Union of Concerned Scientists. "Scientific Integrity in Policy: An Investigation Into the Bush Administration's Misuse of Science." http://www.ucsusa.org/scientific_integrity/abuses_of_science/reports-scientific-integrity.html (accessed August. 20, 2014).

Wikiquote. "John Godfrey Saxe." https://en.wikiquote.org/wiki/John_Godfrey_Saxe (accessed August 9, 2016).

Wikiquote. "William O. Douglas." https://en.wikiquote.org/wiki/William_O._Douglas (accessed August 9, 2016).

Further Reading

Aziz, J. "The World's Dumbest Idea: Taxing Solar Energy." *The Week*, April 17, 2014. http://theweek.com/articles/447732/worlds-dumbest-idea-taxing-solar-energy (accessed October 13, 2015).

Biesecker M., and T. Krisher. "2016 VW Diesels Have New Software Affecting Emissions Tests." *Yahoo! Finance*, October 14, 2015. http://finance.yahoo.com /news/2016-vw-diesels-software-affecting-142019079.html (accessed November 24, 2015).

Bingham, C. "Under Mined: When a Flood of Toxic Mining Sludge Wreaked Havoc in Appalachia, How Did the White House Respond? By Letting the Coal Company Off the Hook and Firing the Whistleblower." *Washington Monthly*, January/ February 2005. http://www.washingtonmonthly.com/features/2005/0501.bingham .html (accessed August 20, 2014).

Bomey, N., and C. Woodyard. "EPA: VW Cheated on Audi, Porsche Diesel SUVs, Too." *USA Today*, November 2, 2015. http://www.usatoday.com/story/money /cars/2015/11/02/epa-diesel-suv-volkswagen-audi-porsche/75044132 (accessed November 24, 2015).

Boston, W. "Volkswagen Emissions Investigation Zeroes In on Two Engineers." *Wall Street Journal*, October 5, 2015. http://www.wsj.com/articles/vw-emissions-probe-zeroes-in-on-two-engineers-1444011602 (accessed November 24, 2015).

Brodwin, E. "There's a Monstrous Issue at the Heart of VW Scandal That No One's Talking About." *Yahoo! Autos*, September 22, 2015. https://www.yahoo.com /autos/s/theres-monstrous-issue-heart-vw-scandal-no-ones-172600434.html ?nf=1 (accessed November 24, 2015).

Christen, K. "EU e-Waste Rules Driving Change in United States." *Environmental Science and Technology* 2003;January 1:13A.

Clinch, M. "Europe's Automakers Caught Up in VW Storm." *Yahoo! Finance*, September 24, 2015. http://finance.yahoo.com/news/bmw-shares-slip-report-high -094815297.html (accessed November 24, 2015).

Cooney, C. M. "President Bush Cuts U.S. EPA Budget." *Environmental Science and Technology* 2005;April 1:148A.

Cooter, W. S. "Clean Water Act Assessment Processes in Relation to Changing U.S. Environmental Protection Agency Management Strategies." *Environmental Science and Technology* 2004;September 14:5265–5272.

Dunnivant, F. M., and E. Anders. *A Basic Introduction to Pollutant Fate and Transport: An Integrated Approach with Chemistry, Modeling, Risk Assessment, and Environmental Legislation* (pp. 337–393). New York: Wiley, 2006.

The Economist. "A Mucky Business," September 26, 2015. http://www.economist .com/news/briefing/21667918-systematic-fraud-worlds-biggest-carmaker-threatens-engulf-entire-industry-and.

EPA. "25 Years of the Safe Drinking Water Act: History and Trends." EPA publication number 816-R-99-007. Washington, DC: Author, December 1999.

Erickson, B. E. "House Clears Chemicals Bill." *Chemical Engineering and News* 2015;June 29:6.

Erickson, B. E. "New Push to Reform Chemical Law." *Chemical Engineering and News* 2015;October 5:24–25.

Exner, J. H. "Science and Policy: Who Speaks for Science?" *Chemical and Engineering News* 2005;December 12:30.

Exner, J. H. "Toward a Global Environmental Ethic." *Chemical and Engineering News* 2005;April 18:49.

Gibb, S. K. "Congress Targets EPA Science Advice." *Chemical Engineering and News* 2015;June 8:23.

Handson, D. J. "Administration Policies Are Panned." *Chemical and Engineering News* 2006;May 8:29–30.

Hogue, C. "Chemicals Management." *Chemical and Engineering News* 2005;September 19:27–28.

Hogue, C. "'Clean Skies' a Dirtier Policy." *Chemical and Engineering News* 2005;January 24:8.

Hogue, C. "Conflict-of-Interest Debate Intensifies: Should Working for Industry Disqualify Scientists from Advising EPA?" *Chemical and Engineering News* 2001;July 30:38–39.

Hogue, C. "Court Orders EPA to Control Toxic Emissions." *Chemical and Engineering News* 2006;August 14:11.

Hogue, C. "EPA Regulation Is Overturned." *Chemical and Engineering News* 2006:March 27:8.

Hogue, C. "Managing Chemicals Globally." *Chemical and Engineering News* 2012;October 8:41.

Hogue, C. "Mixed Reception for Chemicals Bill." *Chemical and Engineering News* 2010;August 2:11.

Hogue, C. "Negotiators Finish Pact on Chemicals." *Chemical and Engineering News* 2006;February 13:12.

Hogue, C. "Plans Pulled." *Chemical and Engineering News* 2005;July 11:10.

Hogue, C. "Regulating Chemicals." *Chemical and Engineering News* 2006;April 10:44–46.

Hogue, C. "Safely Managing Chemicals." *Chemical and Engineering News* 2006;February 27:31–35.

Hogue, C. "States Ascend." *Chemical and Engineering News* 2011;March 21:36–38.

Hogue, C. "States Take the Lead." *Chemical and Engineering News* 2013;February 18:37–39.

Hogue, C. "Updating Chemical Control Law." *Chemical and Engineering News* 2010;March 22:40–41.

Hogue, C. "Violations and Competition." *Chemical and Engineering News* 2004;August 23:26.

Huffington, A. "Good-bye, Mr. President: The Secret Resignation Letters." *Alternet*, December 3, 2003. http://www.alternet.org/story/17309/good-bye,_mr._president %3A_the_secret_resignation_letters (accessed August 20, 2014).

Hyde, J. "Volkswagen Admits 11 Million Vehicles Had Fake Pollution Controls." *Yahoo! Autos*, September 22, 2015. https://www.yahoo.com/autos/volkswagen -admits-11-million-vehicles-had-fake-129635341557.html (accessed November 24, 2015).

Hyde, J. "Why VW Might Not Be the Only Automaker with a Pollution Problem." *Yahoo! Autos*, September 21, 2015. https://www.yahoo.com/autos/why-vw-might -not-be-the-only-automaker-with-a-129578004067.html (accessed November 24, 2015).

IndiaDivine.org. "Bush Administration Plans to Relax Toxic Controls—Again," June 14, 2005. http://www.indiadivine.org/content/topic/1782716-bush-admini stration-plans-to-relax-toxic-controls-again (accessed August 20, 2014).

Kahl, J. S., J. L. Stoddard, R. Haeuber, S. G. Paulsen, R. Birnbaum, F. A. Deviney, et al. "Have U.S. Surface Waters Responded to the 1990 Clean Air Act Amendments?" *Environmental Science and Technology* 2004;December 15:484–490A.

Kennedy, R. F., Jr. "Crimes Against Nature." *Rolling Stone*, December 11, 2003. http:// www.cfr.org/united-states/rolling-stone-crimes-against-nature/p12902   (accessed August 20, 2014).

Kinzig, A. P., P. R. Ehrlich, L. J. Alston, K. Arrow, S. Barrett, T. G. Buchman, et al. "Social Norms and Global Environmental Challenges: The Complex Interaction of Behaviors, Values, and Policy." *Bioscience* 2013;63(3):164–175.

Lazarus, R. J. *The Making of Environmental Law*. Chicago: University of Chicago Press, 2004.

Morgan, M. G. "The U.S. Congress Needs Advice About Science and Technology." *Environmental Science and Technology*. 2004;August 15:306–312A.

Morgan, S. "VW Revs Up Recall Plan, Hunts for Culprits in Pollution Scam." *Yahoo! News*, October 1, 2015. http://news.yahoo.com/vw-says-1-8-mn-commercial -vehicles-emission-092856765.html (accessed November 24, 2015).

Morgan, S. "VW Says Probe Into Pollution Scam to Take Months." *Yahoo! News*, October 1, 2015. http://news.yahoo.com/vw-ex-ceo-not-facing-formal-probe -german-093657842.html (accessed November 24, 2105).

Morse, P. M. "Last-Minute Push for REACH Rule." *Chemical and Engineering News* 2010;November 29:15–17.

New Scientist. "It May Be Bad News, but We Need to Hear It." 2006;February:5.

New Scientist. "Vote for Science." 2006;October 7:6.

New York Times Opinions Page. "Nature at Bay." *New York Times*, May 9, 2005. http://www.nytimes.com/2005/05/09/opinion/09mon1.html (accessed August 20, 2014).

Pearce, F. "Climate Report 'Was Watered Down.'" *New Scientist* 2007; March 10:10.

Rauber, P. "See No Evil: How the White House Edits Out Global Warming." *Sierra* 2006;January/February:36–37. https://vault.sierraclub.org/sierra/200601/decoder .asp (accessed July 13, 2016).

Renner, R. "Florida, EPA Slammed for Regulatory Failures." *Environmental Science and Technology* 2004;August 1:285A.

Repanshek, K. "Bush vs. Science: Round Five Jabs." *Discover* 2006;January:53.

Reuters. "Factbox: Diesel Engines and How VW's 'Defeat Device' Worked," September 29, 2015. http://www.reuters.com/article/2015/09/29/us-usa-volkswagen-diesel -factbox-idUSKCN0RN25M20150929 (accessed November 24, 2015).

Revkin, A. C. "Ex-Bush Aide Who Edited Climate Reports to Join ExxonMobil." *New York Times,* June 15, 2005. http://www.nytimes.com/2005/06/15/science/14cnd -climate.html?_r=0 (accessed August 20, 2014).

Schaeffer, E. "Clearing the Air: Why I Quit Bush's EPA." *Washington Monthly,* July/ August, 2002. http://www.washingtonmonthly.com/features/2001/0207.schaeffer .html (accessed August 20, 2014).

Schnoor, J. L. "Comment: U.S. Is No Longer the Environmental Leader." *Environmental Science and Technology* 2005;May 1:187A.

Schulz, W. G. "Judging Science." *Chemical and Engineering News* 2006;February 27:36.

Shogren, E. "Feds Hope $5 Billion Settlement a Lesson for Polluters." National Public Radio, April 5, 2014. http://www.npr.org/2014/04/05/299204172/feds-hope-5 -billion-settlement-a-lesson-for-polluters (accessed August 20, 2014).

Smith, G. "Volkswagen Hires BP's Deepwater Defense Team as the Lawsuits Start." *Fortune*, September 23, 2015. http://fortune.com/2015/09/23/volkswagen-hires -bps-deepwater-defense-team-as-the-lawsuits-start (accessed November 24, 2015).

Snow, N. "Anadarko Settles Legacy Claims Against Kerr-McGee for $5.15 Billion." *Oil & Gas Journal,* April 7, 2014. http://www.ogj.com/articles/print/volume-112 /issue-4a/general-interest/anadarko-settles-legacy-claims-against-kerr-mcgee -for-5–15-billion.html (accessed August 20, 2014).

Switzer, J. V. *Environmental Politics: Domestic and Global Dimensions,* 4th ed. Belmont, CA: Thomson-Wadsworth, 2004.

Tullo, A. "Regulation: Second Phase of Europe's Chemical Law Closes, with Company Complaints." *Chemical and Engineering News* 2013;June 10:7.

Tutt, P. "Martin Winterkorn Resigns as Volkswagen CEO." CNBC, September 24, 2015. http://www.cnbc.com/2015/09/23/martin-winterkorn-resigns-as-volkswagen -ceo.html (accessed November 24, 2015).

Waxman, H. A. *About Politics and Science: The State of Science Under the Bush Administration.* http://oversight-archive.waxman.house.gov/documents/2008013 0103545.pdf (accessed August 20, 2014).

Weinberg, P., and K. A. Reilly. *Understanding Environmental Law,* 3rd ed. New York: LexisNexis, 2013.

Wikipedia. "American Exceptionalism." http://en.wikipedia.org/wiki/American _exceptionalism (accessed September 27, 2014).

8. The Rapid Advancement of Technology

Notes

Feynman, R. P. http://www.brainyquote.com/quotes/quotes/r/richardpf104988.html (accessed August 9, 2016).

Grossman, L. "Star Power: The Race to Build the World's First Commercial Fusion Reactor Is Heating Up." *Time*, November 2, 2015.

Huesemann, M. H. "Recognizing the Limits of Environmental Science and Technology." *Environmental Science and Technology* 2003;July 1:259A–261A.

Keller, G. M. http://www.azquotes.com/quote/585887 (accessed August 9, 2016).

Mumford, L. http://www.brainyquote.com/quotes/quotes/l/lewismumfo400181 .html (accessed August 9, 2016).

Orwell, G. https://en.wikiquote.org/wiki/George_Orwell (accessed August 9, 2016).

Weisman, Alan. *Countdown: Our Last, Best Hope for a Future on Earth*. Boston: Back Bay Books, 2014.

Further Reading

Bomgardner, M. M. "Fuel Cell Cars Start to Arrive." *Chemical Engineering and News* 2014;November17:17–20.

Bulbs.com. "History of the Light Bulb." http://www.bulbs.com/learning/history .aspx (accessed December 18, 2014).

Cardwell, D. "Solar and Wind Energy Start to Win on Price vs. Conventional Fuels." *New York Times*, November 23, 2014. http://www.nytimes.com/2014/11/24 /business/energy-environment/solar-and-wind-energy-start-to-win-on-price -vs-conventional-fuels.html?_r=0 (accessed October 13, 2015).

Edelstein, S. "The Coming Solar Power Boom: Charts Tell the Story, Grid Parity in 2 Years." *Green Car Reports*, November 13, 2014. http://www.greencarreports .com/news/1095447_the-coming-solar-power-boom-charts-tell-the-story-grid -parity-in-2-years (accessed October13, 2015).

Edelstein, S. "Graphene May Double Solar Cell Power, Harvest Hydrogen from Air." *Green Car Reports*, December 30, 2014. http://www.greencarreports.com/news /1096080_graphene-may-double-solar-cell-power-harvest-hydrogen-from-air (accessed October 13, 2015).

Jacoby, M. "Unleashing the Power of Thorium." *Chemical and Engineering News* 2015;July 6:44–46.

Koch, R. "The End of Capitalism." *Huffington Post,* January 14, 2014. http://www .huffingtonpost.com/richard-koch/the-end-of-capitalism_b_4593969.html (accessed December 30, 2014).

Kuittine, T. "A Lamp That Lasts 40 Years Will Debut in May." *BGR News,* December 30, 2014. https://bgr.com/2014/12/30/jake-dyson-aerial-light (accessed October 13, 2015).

McGrath, M. "'Skunk Power' Creates Confusion Over Nuclear Fusion." *BBC News: Science and Environment,* November 16, 2014. http://www.bbc.com/news/science -environment-29710811 (accessed December 18, 2014).

Mearian, L. "Rooftop Solar Electricity on Pace to Beat Coal, Oil." *ComputerWorld,* November 18, 2014. http://www.computerworld.com/article/2848875/rooftop -solar-electricity-on-pace-to-beat-coal-oil.html (accessed October 13, 2015).

Moritsugu, K. "Toyota to Start Sales of Fuel Cell Car Next Month." *Yahoo! Tech,* November 18, 2014. https://www.yahoo.com/tech/s/toyota-launch-fuel-cell-car -022146611.html?nf=1 (accessed October 13, 2015).

Orwig, J. "Germany Is About to Start Up a Monster Machine That Could Revolutionize the Way We Use Energy." *Yahoo! Finance,* October 30, 2015. http:// finance.yahoo.com/news/germany-start-monster-machine-could-152111129 .html (accessed November 25, 2015).

Plumer, B. "The Cost of Wind and Solar Power Keeps Dropping All Over the World." *Vox,* February 5, 2015. http://www.vox.com/2015/2/5/7977869/wind-solar-costs -falling (accessed October 13, 2015).

Population. "A Series of Papers on Population." *Science* 2011;333(6042):538–592.

Roberts, D. "Here's What It Would Take for the US to Run on 100% Renewable Energy." *Vox,* June 9, 2015. http://www.vox.com/2015/6/9/8748081/us-100-percent -renewable-energy (accessed August 31, 2015).

Satell, G. "How the Energy Revolution Will Transform How We Live and Work." *Forbes,* February 8, 2015. http://www.forbes.com/sites/gregsatell/2015/02/08/how -the-energy-revolution-will-transform-how-we-live-and-work/#35768e 0624c8 (accessed September 4, 2016).

Scott, A. "Challenging Lithium-Ion Batteries." *Chemical Engineering and News* 2015;July 20:18–19.

Shalal, A. "Lockheed Makes Breakthrough on Fusion Energy Project." *Yahoo! News,* October 15, 2014. https://www.yahoo.com/news/lockheed-says-makes-break through-fusion-energy-project-123840986--finance.html?ref=gs.

Talbot, D. "Does Lockheed Martin Really Have a Breakthrough Fusion Machine?" *MIT Technology Review,* October 20, 2014. http://www.technologyreview.com /news/531836/does-lockheed-martin-really-have-a-breakthrough-fusion -machine (accessed December 18, 2014).

Voelcker, J. "Nickel-Metal-Hydride Batteries for Electric Cars? Energy Density Can Rise 10-fold: Researchers." *Green Car Reports,* March 2, 2015. http://www.greencar

reports.com/news/1097016_nickel-metal-hydride-batteries-for-electric-cars
-energy-density-can-rise-10-fold-researchers (accessed October 13, 2015).

Wikipedia. "Light-emitting Diodes." http://en.wikipedia.org/wiki/Light-emitting
_diode (accessed December 18, 2014).

Wikipedia. "Moore's Law." http://en.wikipedia.org/wiki/Moore%27s_law (accessed
December 18, 2014).

Wynn, G. "Solar Closing In on Cost of Coal-Fired Power." *Climate Change News*,
March 5, 2015. http://www.climatechangenews.com/2015/03/05/solar-closing
-in-on-cost-of-coal-fired-power-deutsche-bank (accessed October 13, 2015).

9. Humans' Greatest Challenge

Notes

Abrams, L. "House Republicans Just Passed a Bill Forbidding Scientists from Advising
the EPA on Their Own Research." *Salon*, November 19, 2014. http://www.salon
.com/2014/11/19/house_republicans_just_passed_a_bill_forbidding_scientists
_from_advising_the_epa_on_their_own_research (accessed November 25, 2014).

Arrhenius, S. "On the Influence of Carbonic Acid in the Air Upon the Temperature
of the Ground." *Philosophical Magazine and Journal of Science* 1896;Series 5, Volume 41:237–276.

Associated Press. "What US, China Agreed to Do to Limit Emissions." *Yahoo! Finance*, November 12, 2014. http://finance.yahoo.com/news/us-china-pledging
-limit-emissions-094031594.html (accessed November 25, 2014).

Atkin, E. "Here's Who to Blame for America's Increased Contribution to Global
Warming in 2013." ThinkProgress.org, October 1, 2014. http://thinkprogress
.org/climate/2014/10/01/3574709/united-states-industry-global-carbon
-emissions-2013 (accessed November 25, 2014).

Boehner, J. "Boehner on President Obama's National Energy Tax." Speaker.org, August 3, 2015. http://www.speaker.gov/press-release/boehner-responds-president
-obama-s-national-energy-tax (accessed August 16, 2016).

Borenstein, S. "Arctic Ice Shrinks to All-Time Low; Half 1980 Size." *Yahoo! News*,
September 19, 2012. http://news.yahoo.com/arctic-ice-shrinks-time-low-half
-1980-size-175242723.html (accessed August 15, 2016).

Borenstein, S. "Global Warming Dials Up Our Risks, UN Report Says." Associated
Press, March 30, 2014. http://bigstory.ap.org/article/global-warming-dials-our
-risks-un-report-says (accessed August 7, 2014).

Bump, P. "The Main Climate Change Question: When Do You Want to Pay for It,
and How?" *The Wire*, April 14, 2014. http://www.thewire.com/politics/2014/04
/the-main-climate-change-question-when-do-you-want-to-pay-for-it-and-how
/360621 (accessed August 7, 2014).

Cama, T. "McConnell: EPA Proposal 'Dagger in the Heart' of Middle Class." *The Hill,* June 6, 2014. http://thehill.com/policy/energy-environment/207897-mcconnell -epa-proposal-a-dagger-in-the-heart-of-middle-class (accessed August 9, 2016).

Carey, J. "Climate Wars: Episode Two." *Business Week* 2007;April 23:90–92. http:// www.businessweek.com/stories/2007-04-22/climate-wars-episode-two (accessed August 11, 2014).

EPA. "Greenhouse Gas Reporting Program (GHGRP)." http://www.epa.gov /ghgreporting/ghgdata/reported/index.html (accessed September 1, 2016).

Fischetti, M. "29 Bullets Tell All About Climate Challenge." *Scientific American Blogs,* November 3, 2014. http://environmentalfuture.org/29-bullets-tell-all -about-climate-challenge (accessed November 25, 2014).

Forsythe, M. "Investigating Family's Wealth, China's Leader Signals a Change." *New York Times,* April 20, 2014. http://www.nytimes.com/2014/04/20/world/asia /severing-a-familys-ties-chinas-president-signals-a-change.html.

Gillis, J. "Heat-Trapping Gas Passes Milestone, Raising Fears." *New York Times,* May 10, 2013. http://www.nytimes.com/2013/05/11/science/earth/carbon-dioxide -level-passes-long-feared-milestone.html?pagewanted=all&_r=0 (accessed July 7, 2014).

Gore, A. http://www.notable-quotes.com/g/gore_al.html (accessed August 9, 2016).

Gore, A. "The Turning Point: New Hope for the Climate." *Rolling Stone,* June 18, 2014. http://www.rollingstone.com/politics/news/the-turning-point-new-hope -for-the-climate-20140618 (accessed August 9, 2016).

Harvey, C. "Here's How Little Time We Have Until Global Warming Is Out of Control." *Business Insider,* September 13, 2014. http://www.businessinsider.com/when -will-climate-change-be-out-of-control-2014-9 (accessed November 25, 2014).

Hogue, C. "CO_2 Emissions: Global Releases in 2011 Set a Record High at 31.6 Billion Metric Tons." *Chemical Engineering and News* 2012;June 14:8.

Intergovernmental Panel on Climate Change (ICPP). "Climate Change 2014 Synthesis Report." http://www.ipcc.ch/pdf/assessment-report/ar5/syr/SYR_AR5_FINAL _full_wcover.pdf (accessed August 9, 2016).

Johnson, J. "Energy's Hidden Cost." *Chemical Engineering and News* 2009b;November 2:24–25.

Johnson, J. "Fossil-Fuel Costs." *Chemical Engineering and News* 2009c; October 26:6.

Johnson, J. "Of Climate and National Security." *Chemical Engineering and News* 2012;November 19:10.

Keeling, C. D., R. B. Bacastow, A. E. Bainbridge, C. A. Ekdahl, R. Guenther, and L. S. Waterman. "Atmospheric Carbon Dioxide Variations at Mauna Loa Observatory, Hawaii." *Tellus* 1976;28:538–551.

Mann, M. E. *The Hockey Stick and the Climate Wars: Dispatches from the Front Lines.* New York: Columbia University Press, 2013.

Markowitz, G., and D. Rosner. *Deceit and Denial: The Deadly Politics of Industrial Pollution.* Berkeley: University of California Press, 2013.

McKibben, B. *The End of Nature.* New York: Random House, 2006.

Oreskes, N., and E. M. Conway. *Merchants of Doubt*. New York: Bloomsbury Press, 2010.

Osborne, H. "State of the Union Address: 10 Quotes on Climate Change from Barack Obama." *International Business Times*, January 21, 2015. http://www.ibtimes.co.uk/state-union-address-10-quotes-climate-change-barack-obama-1484442 (accessed August 9, 2016).

Parry, W. "Oceans Turning Acidic Faster than Past 300 Million Years." *Live Science*, March 2, 2012. http://www.livescience.com/18786-ocean-acidification-extinction.html (accessed August 7, 2014).

Ritter, K. "Cost of Fighting Warming 'Modest,' Says UN Panel." *Yahoo! News*, April 13, 2014. http://news.yahoo.com/cost-fighting-warming-modest-says-un-panel-153929628--finance.html (accessed August 7, 2014).

Thoning, K. W. P. Tans, and W. D. Komhyr. "Atmospheric Carbon Dioxide at Mauna Loa Observatory 2. Analysis of the NOAA GMCC Data, 1974–1985." *Journal of Geophysical Research* 1989;94:8549–8565.

Further Reading

American Chemical Society. *Chemistry in Context: Applying Chemistry to Society*, 6th ed. (pp. 100–149). New York: McGraw-Hill, 2009.

Ashtari, S. "We've Reached the Point Where Climate Change Deniers Need to Be Reminded That It Snows Every Year." *Huffington Post*, December 11, 2013. http://www.huffingtonpost.com/2013/12/11/jared-huffman-climate-change_n_4426228.html (accessed August 18, 2014).

Associated Press. "China to Ban All Coal Use in Beijing by 2020." *Yahoo! News*, August 5, 2014. http://www.yahoo.com/news/china-ban-coal-beijing-2020-083022870.html (accessed August 15, 2014).

Associated Press. "What US, China Agreed to Do to Limit Emissions." *Yahoo! Finance*, November 12, 2014. http://finance.yahoo.com/news/us-china-pledging-limit-emissions-094031594.html (accessed August 16, 2016).

Baetz, J., and J. McDonald. "Obama's Emissions Plan Could Boost Climate Talks." Associated Press, June 3, 2014. http://bigstory.ap.org/article/obama-plan-might-help-boost-climate-talks (accessed August 18, 2014).

Bailey, R. A., H. M. Clark, J. P Ferris, S. Krause, and R. L. Strong. *Chemistry of the Environment* (pp. 39–70). New York: Academic Press, 2002.

Baird, C. *Environmental Chemistry* (pp. 149–190). New York: Freeman, 1995.

Bakewell, S. "China Power Seen Doubling with Renewables as Coal Holds Sway." *Bloomberg*, August 26, 3013. http://www.bloomberg.com/news/2013-08-27/china-power-seen-doubling-with-renewables-as-coal-holds-sway.html (accessed August 18, 2014).

Barry, C., and J. Wardell. "Obama, G-8 Leaders Agree on Climate Target." *Huffington Post*, July 8, 2009. http://www.huffingtonpost.com/2009/07/08/obama-g8-leaders-agree-on_n_227925.html (accessed August 12, 2014).

Baum, R. M. "Climate-Change News." *Chemical Engineering and News* 2009;June 22:3.

Baum, R. M. "Global Warming Denial." *Chemical Engineering and News* 2007;February 12:3.

Baum, R. M. "Michael Mann's Hockey Stick." *Chemical Engineering and News* 2012;December 10:52.

Bloomberg. "The Future of China's Power Sector," August 27, 2013. http://about.bnef .com/white-papers/the-future-of-chinas-power-sector (accessed August 18, 2014).

Brown, P. "Ozone Layer Most Fragile on Record." *Guardian,* April 27, 2005. http://www.theguardian.com/science/2005/apr/27/environment.research (accessed July 30, 2014).

Budd, J. "Hot and Bothered." *The Economist* 2015;November 28:3–16.

Bump, P. "Climate Change: More Violence, Less Food, and Embarrassment for Political Leaders." *Yahoo! News,* March 31, 2014. http://news.yahoo.com/climate -change-more-violence-less-food-embarrassment-political-142241742.html (accessed August 7, 2014).

Busbee, J. "Chinese Media: Smog Has at Least Five Benefits." *Yahoo! News,* December 10, 2013. http://news.yahoo.com/chinese-government-tries-to-spin-smog -as-a-healthy-benefit--and-aid-to-national-defense-174649885.html (accessed August 18, 2014).

Cappiello, D. "Obama Takes On Coal with First-Ever Carbon Limits." Associated Press, September 20, 2014. http://bigstory.ap.org/article/obama-takes-coal-first -ever-carbon-limits (accessed August 18, 2014).

Cappiello, D., and A. Beam. "Power Plant Plan Further Clouds Coal's Future." Associated Press, June 3, 2014. http://bigstory.ap.org/article/power-plant-plan-further -clouds-coals-future (accessed August 18, 2014).

Cappiello, D., and J. Lederman. "Obama Carbon Rule Gives States More Time to Comply." *Yahoo! News,* June 2, 2014. http://news.yahoo.com/obama-carbon-rule -gives-states-more-time-comply-131341354--finance.html (accessed August 18, 2014).

Cardwell, D. "Solar and Wind Energy Start to Win on Price vs. Conventional Fuels." *New York Times,* November 23, 2014. http://www.nytimes.com/2014/11/24 /business/energy-environment/solar-and-wind-energy-start-to-win-on-price -vs-conventional-fuels.html?_r=1 (accessed November 25, 2014).

Chemical Engineering and News. "Court Says States May Regulate Greenhouse Gases." 2007;September 17:27.

Chemical Engineering and News. "Delaying Emissions Cuts Has High Costs." 2013;November 11:22.

Chemical Engineering and News. "EPA Is Sued Over Carbon Emissions." 2006;May 8:28.

Chestney, N. "Global Carbon Emissions Hit Record High in 2012." Reuters, June 10, 2013. http://www.reuters.com/article/2013/06/10/us-iea-emissions -idUSBRE95908S20130610 (accessed August 11, 2014).

Childers, A. "Supreme Court Could Clarify EPA Authority to Regulate Carbon Pollution." *Bloomberg*, February 20, 2014. http://www.bloomberg.com/news/2014-02-20/supreme-court-could-clarify-epa-authority-to-regulate-carbon-pollution.html (accessed August 18, 2014).

Chow, D. "5 Places Already Feeling the Effects of Climate Change." *Live Science*, November 22, 2013. http://www.livescience.com/41380-climate-change-places-at-risk.html (accessed August 18, 2014).

Clayton, M. "U.S. Coal Power Boon Suddenly Wanes." *Christian Science Monitor*, March 4, 2008. http://www.csmonitor.com/Business/2008/0304/p01s07-usec.html (accessed August 7, 2014).

Cooke, K. "China's 'War on Pollution' Helps Kick Coal Habit." *EcoWatch*, September 5, 2014. http://ecowatch.com/2014/09/05/china-war-pollution-kick-coal-habit (accessed November 25, 2014).

Doney, S. C. "The Dangers of Ocean Acidification." *Scientific American* 2006; March:58–65.

Dow, K, and T. E. Downing. *The Atlas of Climate Change: Mapping the World's Greatest Challenge*. London: Earthscan, 2006.

Doyle, A. "Methane Bubbles in Arctic Seas Stir Warming Fears." Reuters, March 4, 2010. http://uk.reuters.com/article/2010/03/04/us-climate-methane-idUKTRE6233ZU20100304 (accessed August 11, 2014).

Doyle, A. "Tipping Point: On Horizon for Greenland Ice." Reuters, February 4, 2008. http://www.reuters.com/article/2008/02/04/us-climate-tipping-idUSL01827 6620080204 (accessed August 11, 2014).

Drajem, M. "Obama Said to Ban New Coal Plants Without Carbon Controls." *Bloomberg*, September 11, 2013. http://nacbe.com/obama-said-to-ban-new-coal-plants-without-carbon-controls-bloomberg (accessed August 18, 2014).

EEnergy Informer. "Half New US Capacity Renewable in 2014." BreakingEnergy.com, March 5, 2015. http://breakingenergy.com/2015/03/05/half-new-us-capacity-renewable-in-2014 (accessed October 13, 2015).

Ekwuzel, B. "Climate Disinformation Continues to Harm U.S. Communities." *Live Science*, August 22, 2013. http://www.livescience.com/39005-climate-disinformation-harming-us-communities.html (accessed August 18, 2014).

Farrell, P. B. "Historic U.S.-China Climate Deal Has GOP 'Freaking Out.'" *Market Watch*, November 17, 2014. http://www.marketwatch.com/story/historic-us-china-climate-deal-has-gop-freaking-out-2014-11-15?siteid=yhoof2 (accessed November 25, 2014).

Foley, S. "Tom Steyer Challenges David Koch on Climate Change Stance." *Financial Times*, February 2, 2015.

Freedman, A. "Why Al Gore Is Hopeful on Global Warming." *Mashable*, September 18, 2014. http://mashable.com/2014/09/18/al-gore-is-hopeful-global-warming (accessed August 31, 2015).

Fritz, A. "Summer 2014 Was Record Warmest on Earth, Says NOAA." *Washington Post*, September 18, 2014. http://www.washingtonpost.com/blogs/capital

-weather-gang/wp/2014/09/18/june-july-august-set-record-for-warmest
-summer-on-earth-says-noaa (accessed November 25, 2014).

Gardner, T. "EPA Cleared to Regulate U.S. Emissions as Congress Stalls." Reuters, December 7, 2009. http://www.reuters.com/article/2009/12/07/us-usa-climate -epa-idUSTRE5B628820091207 (accessed August 7, 2014).

Geman, B. "EPA Is on a Legal Winning Streak." *National Journal,* April 29, 2014. http://www.nationaljournal.com/energy/epa-is-on-a-legal-winning-streak -20140429 (accessed August 7, 2014).

Gibb, S. K. "China Emission Cuts." *Chemical Engineering and News* 2015;April 27:22–23.

Gibb, S. K. "Natural Gas Overtakes Coal as U.S. Electric Source." *Chemical Engineering and News* 2015;July 20:24.

Gillis, J. "2012 Was Hottest Ever in U.S." *New York Times,* January 18, 2013. http:// www.nytimes.com/2013/01/09/science/earth/2012-was-hottest-year-ever-in -us.html.

Girard, J. E. *Principles of Environmental Chemistry* (pp. 66–91). Sudbury, MA: Jones and Bartlett, 2010.

Hackmann, H., S. C. Moser, and A. L. St. Clair. "The Social Heart of Global Environmental Change." *Nature Climate Change* 2014;4:653–655.

Hanley, C. J. "Global Experts: Warming Could Double Food Prices." Phys.org, December 1, 2010. http://phys.org/news/2010-12-global-experts-food-prices.html (accessed August 11, 2014).

Hanley, C. J. "Mass Migrations and War: Dire Climate Scenario." CommonDreams .org, February 22, 2009. http://www.commondreams.org/news/2009/02/22 /mass-migrations-and-war-dire-climate-scenario (accessed August 11, 2014).

Hansen, J. *Storms of My Grandchildren: The Truth About the Coming Climate Catastrophe and Our Last Chance to Save Humanity.* New York: Bloomsbury, 2009.

Harvey, C. "We Are Pumping Carbon Into the Atmosphere at Increasing Rates—2013 Set a New Record Jump." *Business Insider,* September 21, 2014. http://finance .yahoo.com/news/pumping-carbon-atmosphere-increasing-rates-163900943 .html (accessed November 25, 2014).

Hileman, B. "Climate Change Threatens Global Economy." *Chemical Engineering and News* 2006;November 6:7.

Hileman, B. "Ice Core Record Extended." *Chemical Engineering and News* 2005;November 28:7.

Hileman, B. "Stark Effects from Global Warming." *Chemical Engineering and News* 2005;March 21:47–48.

Hileman, B. "States Act to Protect Climate." *Chemical Engineering and News* 2005;July 11:24–26.

Hileman, B. "UN Issues Climate Warnings." *Chemical Engineering and News* 2007;February 12:17.

Hogue, C. "Climate-Change Confab." *Chemical Engineering and News* 2013;November 11:23–25.

Hogue, C. "Climate Rule Under Fire." *Chemical and Engineering News* 2015;November 9:8.

Hogue, C. "CO_2 Levels Climb Ever Higher." *Chemical and Engineering News* 2015:November 16:6.

Hogue, C. "Court Affirms Regulation of Greenhouse Gases." *Chemical Engineering and News* 2012;July 2:19.

Hogue, C. "EPA Actions Under Fire." *Chemical Engineering and News* 2011;February 4:5.

Hogue, C. "How Emissions Offsets Work." *Chemical Engineering and News* 2009;March 23:33–34.

Hogue, C. "Reining in Brown Clouds." *Chemical Engineering and News* 2009;August 24:26–27.

Hogue, C., and J. Johnson. "House Panel Votes to Halt Climate Rules: Republicans Push Antiregulatory Agenda." *Chemical Engineering and News* 2011;March 14:9.

Hood, M. "Sea Levels to Surge 'at Least a Metre' by Century End." *Free Republic*, March 11, 2009. http://www.freerepublic.com/focus/news/2204077/posts (accessed August 11, 2009).

Hull, R. B. "God's Will and the Climate." *New Scientist* 2006;April 1:23.

Johnson, J. "Action on CO_2 Emissions." *Chemical Engineering and News* 2013;September 30:8.

Johnson, J. "Angst Over Cap and Trade." *Chemical Engineering and News* 2009;June 15:22.

Johnson, J. "Big Savings Through Efficiency." *Chemical Engineering and News* 2006;October 9:28–29.

Johnson, J. "Changes Ahead for Old Power Plants." *Chemical Engineering and News* 2011;September 19:22–24.

Johnson, J. "Climate-Change Solutions." *Chemical Engineering and News* 2007;May 14:10.

Johnson, J. "Coal Plant Permit Blocked." *Chemical Engineering and News* 2008;November 24:9.

Johnson, J. "EPA Proposes CO_2 Cuts for Power Plants." *Chemical Engineering and News* 2014;June 9:7.

Johnson, J. "Fixing Global Warming." *Chemical Engineering and News* 2007;February 19:11.

Johnson, J. "The Foul Side of 'Clean Coal.'" *Chemical Engineering and News* 2009;February 23:44–47.

Johnson, J. "Methane's Role in Climate Change." *Chemical Engineering and News* 2014;July 7:9–15.

Johnson, J. "The Natural Gas Advantage." *Chemical Engineering and News* 2013;June. 24:29.

Johnson, J. "The New King of Fossil Fuel." *Chemical Engineering and News* 2013;October 21:27.53.

Johnson, J. "Regulating CO_2 Emissions." *Chemical Engineering and News* 2013;October 21:8.

Johnson, J. "Stuck on Fossil Fuels." *Chemical Engineering and News* 2012;June 4:27.

Johnson, J. "Stumbling on the Path to 'Clean Coal.'" *Chemical Engineering and News* 2012;July 16:37–39.

Johnson, J. "Top Brass Fear Climate Change." *Chemical Engineering and News* 2007;April 23:8.

Johnson, J. "Using CO_2 as a Raw Material." *Chemical Engineering and News* 2014;November 17:26–27.

Johnson, J., and A. Scott. "Carbon Conundrum." *Chemical Engineering and News* 2013;February 18:12–25.

Jordans, F. "Climate Meeting to Discuss Future of Fossil Fuels." *Huffington Post*, April 5, 2014. http://www.huffingtonpost.com/2014/04/05/ipcc-meeting-berlin -fossil-fuels_n_5096948.html (accessed August 7, 2014).

Kurtenbach, E. "Costs of Climate Change Steep but Tough to Tally." Associated Press, March 31, 2014. http://bigstory.ap.org/article/costs-climate-change-steep -tough-tally (accessed August 18, 2014).

Lavelle, J. "Additive Curbs Bovine Methane." *Chemical Engineering and News* 2015;August 17:8.

Lederman, J., and M. Daly. "Power Plant Limits at Center of Obama Climate Plan." *Yahoo! News*, June 25, 2013. http://news.yahoo.com/power-plant-limits-center -obama-100316174.html (accessed August 11, 2014).

Litterman, B. "Time Is Key to Putting a Price on Climate Risk." *Live Science*, September 16, 2013. http://www.livescience.com/39572-how-quickly-can-society -respond-to-global-risk.html (accessed August 18, 2014).

Live Science. "Greenhouse Gases Hit Record High in 2011," November 20, 2012. http://www.livescience.com/24919-greenhouse-gases-record-high.html (accessed August 11, 2014).

Lobello, C. "China's Newest Environmental Disaster." *The Week*, September 26, 2014. http://theweek.com/article/index/250284/chinas-newest-environmental -disaster (accessed August 18, 2014).

LoGiurato, B. "Boehner: Obama's Plan Is 'Nuts.'" *Business Insider*, June 2, 2014. http://www.businessinsider.com/boehner-obama-carbon-plan-nuts-2014-6 (accessed August 18, 2014).

LoGiurato, B. "Obama Is Proposing the Most Sweeping Move Yet to Combat Global Warming, and It's Already Controversial." *Business Insider*, June 2, 2014. http:// www.businessinsider.com/obama-climate-change-regulations-coal-plants -2014-6 (accessed August 18, 2014).

Lovan, D. "Kentucky Plant Emblematic of Move from Coal to Gas." Associated Press, June 4, 2014. http://bigstory.ap.org/article/kentucky-plant-emblematic -move-coal-gas (accessed August 18, 2014).

Lynas, M. *Six Degrees: Our Future on a Hotter Planet.* New York: Fourth Estate, 2007.

Mack, E. "Forget 'On Record,' 2014 May Has Been Warmest Year in Last 2,000." *Forbes*, January 18, 2015. http://www.forbes.com/sites/ericmack/2015/01/18 /forget-on-record-2014-may-have-been-warmest-year-in-last-5000 (accessed August 31, 2015).

Main, D. "Arctic Temperatures Highest in at Least 44,000 Years." *Live Science*, October 24, 2013. http://www.livescience.com/40676-arctic-temperatures-record -high.html (accessed August 18, 2014).

Mann, M. "Climate-Change Deniers Must Stop Distorting the Evidence." *Live Science*, September 30, 2014. http://www.livescience.com/39957-climate-change -deniers-must-stop-distorting-the-evidence.html (accessed August 18, 2014).

Martelle, S. "King Coal in Court." *Sierra*, 2009;May/June:36–43. http://vault .sierraclub.org/sierra/200905/kingcoal.aspx (accessed August 11, 2014).

Maslin, M. *Global Warming: A Very Short Introduction*. Oxford: Oxford University Press, 2004.

Meadows, D., J. Randers, and D. Meadows. *Limits to Growth: The 30-Year Update*. White River Junction, VT: Chelsea Green, 2004.

Mearian, L. "Rooftop Solar Electricity on Pace to Beat Coal, Oil." *Computerworld*, November 18, 2014. http://www.computerworld.com/article/2848875 /rooftop-solar-electricity-on-pace-to-beat-coal-oil.html (accessed November 25, 2014).

Merchant, B. "Big Oil Says Solar Power Will Win Out." *Motherboard*, October 2, 2013. http://motherboard.vice.com/blog/big-oil-solar-power-will-win-out (accessed August 18, 2014).

Miles, T. "Greenhouse Gas Volumes Reached New High in 2012: WHO." Reuters, November 6, 2013. http://uk.reuters.com/article/2013/11/06/us-greenhouse -idUKBRE9A50ED20131106 (accessed August 18, 2014).

Monbiot, G. "The Denial Industry." *Guardian*, September 19, 2006. http://www .theguardian.com/environment/2006/sep/19/ethicalliving.g2 (accessed August 11, 2014).

Oskin, B. "Africa's Worst Drought Tied to West's Pollution." *Live Science*, June 7, 2013. http://www.livescience.com/37282-north-america-pollution-caused-africa -drought.html (accessed August 11, 2014).

Oskin, B. "2014 Was Earth's Hottest Year on Record." *Live Science*, January 19, 2015. http://www.livescience.com/49479-2014-earths-hottest-year.html (accessed August 31, 2015).

Pacala, S., and R. Socolow. "Stabilization Wedges: Solving the Climate Problem for the Next 50 Years with Current Technologies." *Science* 2004;305:968–972.

Pappas, S. "Global Biodiversity Down 30 Percent in 40 Years." *Live Science*, May 14, 2012. http://www.livescience.com/20307-biodiversity-natural-resources.html (accessed August 7, 2014).

Parkinson, G. "Solar's Insane Cost Drop." *CleanTechnica*, April 16, 2014. http:// cleantechnica.com/2014/04/16/solars-dramatic-cost-fall-may-herald-energy -price-deflation (accessed August 18, 2014).

Pearce, F. "Act Now Before It's Too Late." *New Scientist.* 2005;February 12:8. http://www.newscientist.com/article/mg18524864.300-climate-change-act-now-before-it-is-too-late.html (accessed August 11, 2014).

Pearce, F. "But Here's What They Don't Tell Us." *New Scientist* 2007;February 10. http://www.newscientist.com (accessed August 7, 2014).

Pearce, F. "One Degree and We're Done For." *New Scientist,* 2006;September 30:8–9. http://www.newscientist.com/article/mg19125713.300-climate-change-one-degree-and-were-done-for.html (accessed August 11, 2014).

Pelley, J. "Abrupt Climate Change Threatens National Security." *Environmental Science and Technology.* 2004;May 15:179A.

Pelley, J. "States Take Lead on Climate Change Laws." *Environmental Science and Technology.* 2004;January 15:30A.

Pfeiffer, C. "What the World Would Look Like If All the Ice Melted." *Yahoo! News,* November 5, 2013. http://news.yahoo.com/blogs/the-sideshow/what-the-world-would-look-like-if-all-the-ice-melted-015759322.html (accessed August 18, 2014).

Phillips, M. "The Supreme Court Dims the Lights on Coal Power." *Business Week,* April 29, 2014. http://www.businessweek.com/articles/2014-04-29/the-supreme-courts-epa-ruling-dims-the-lights-on-coal-power-plants (accessed August 7, 2014).

Plautz, J. "Obama's EPA Notches Big Win in Court on Power-Plant Rule." *National Journal,* April 15, 2014. http://www.nationaljournal.com/energy/obama-s-epa-notches-big-win-in-court-on-power-plant-rule-20140415 (accessed August 7, 2014).

Randers, J. *A Global Forecast for the Next Forty Years.* White River Junction, VT: Chelsea Green, 2012.

Reisch, M. S. "Coal's Enduring Power." *Chemical Engineering and News* 2012;November 19:12–17.

Reuters. "U.S. EPA Sues OG&E Over Work at Oklahoma Coal Power Plants," July 9, 2013. http://www.reuters.com/article/2013/07/09/utilities-oge-epa-coal-idUSL1N0FF0PO20130709 (accessed August 18, 2014).

Rood, R. B. "Let's Call It: 30 Years of Above-Average Temperature Means the Climate Has Changed." TheConversation.com, February 26, 2015. http://theconversation.com/lets-call-it-30-years-of-above-average-temperatures-means-the-climate-has-changed-36175 (accessed August 31, 2015).

Saleska, S. "Wake Up, Time for Action." *New Scientist,* 2006;December 9:22.

Scott, A. "Learning to Love CO_2." *Chemical and Engineering News* 2015;November 16:8–16.

Stableford, D. "When Will Climate Change Strike You?" *Yahoo! News,* October 10, 2013. http://news.yahoo.com/when-will-climate-change-strike-135726362.html (accessed August 18, 2014).

Stecker, T. "How Obama's EPA Will Cut Coal Pollution." *Scientific American,* May 30, 2014. https://www.scientificamerican.com/article/how-obama-s-epa-will-cut-coal-pollution (accessed August 18, 2014).

Suzuki, D. "They're Welcome in My Backyard." *New Scientist* 2005;April 16. https://www.newscientist.com/article/mg18624956-400-the-beauty-of-wind-farms (accessed August 11, 2014).

Task, A. "Clean Coal: 'A Terrible Idea Whose Time Has Come,' WIRED's Mann Says." *Daily Ticker,* March 28, 2014. http://finance.yahoo.com/blogs/daily-ticker/clean-coal—a-terrible-idea-whose-time-has-come--wired-s-mann-says-1431 48359.html (accessed August 7, 2014).

Thompson, A. "Climate Scientist: 2 Degrees of Warming Too Much." *Live Science,* December 4, 2013. http://www.livescience.com/41690-2-degrees-of-warming-too -much.html (accessed August 31, 2015).

Toor, A. "Dried Up: Climate Change Could Leave Another Billion People Without Enough Water." *The Verge,* October 11, 2013. http://www.theverge.com/2013/10/11/4828128/water-scarcity-crisis-for-us-fueled-by-climate-change (accessed August 18, 2014).

Torrice, M. "The Limits of Removing Carbon." *Chemical Engineering and News* 2015;August 17:4.

Victor, D. "Climate Change: Embed the Social Sciences in Climate Policy." *Nature Climate Change* 2015;520:27–29.

Walker, M. "A Nation Struggling to Catch Its Breath." *New Scientist,* April 29, 2006. http://www.newscientist.com/article/dn9082-china-struggling-to-catch-its-breath.html (accessed August 11, 2014).

Weart, S. R. *The Discovery of Global Warming.* Cambridge, MA: Harvard University Press, 2003.

Weaver, C. P., S. Mooney, D. Allen, N. Beller-Simms, T. Fish, A. E. Gramsch, et al. "From Global Change Science to Action with Social Sciences." *Nature Climate Change* 2014;4:656–659.

The Week. "Climate Change Report: Things Are Bad, and They're Going to Get Worse," March 31, 2014. http://theweek.com/speedreads/index/259020/speedreads -climate-change-report-things-are-bad-and-theyre-going-to-get-worse (accessed August 18, 2014).

The Week. "Germany Now Gets 28.5 Percent of Its Total Energy from Renewables," July 30, 2014. http://theweek.com/speedreads/index/265542/speedreads-germany-now-gets-285-percent-of-its-total-energy-from-renewables (accessed August 18, 2014).

Werner, C. "To Paris with Love: Stop Fossil Fuel Subsidies." *The Hill,* November 30, 2015. http://thehill.com/blogs/congress-blog/energy-environment/261335-to -paris-with-love-stop-fossil-fuel-subsidies (accessed December 1, 2015).

Wilson, E. "Epic Methane Leak Eases Up." *Chemical and Engineering News* 2016;February 1:35.

Woody, T. "Forget Fracking—Wind Energy Prices Hit a Record Low in the U.S." TakePart.com, August 18, 2014. http://www.takepart.com/article/2014/08/18 /forget-fracking-wind-energy-prices-have-hit-record-low (accessed August 18, 2014).

Woody, T. "Here's Why Al Gore Is Optimistic About the Fight Against Climate Change." *Yahoo! News,* July 7, 2014. http://news.yahoo.com/heres-why-al-gore-optimistic -fight-against-climate-211836883.html?ref=gs (accessed August 31, 2015).

10. Conclusion and Transition to a Bright Future

Notes

Hansen, J. *Storms of My Grandchildren: The Truth About the Coming Climate Catastrophe and Our Last Chance to Save Humanity.* New York: Bloomsbury, 2009.

Lovelock, J. *A Rough Ride to the Future.* New York: Overlook, 2015.

Lynas, M. *Six Degrees: Our Future on a Hotter Planet.* New York: HarperCollins, 2007.

Markowitz, G., and D. Rosner. *Deceit and Denial: The Deadly Politics of Industrial Pollution.* Berkeley: University of California Press, 2013.

Meadows, D., J. Randers, and D. Meadows. *Limits to Growth: The 30-Year Update.* White River Junction, VT: Chelsea Green, 2004.

Oreskes, N., and E. M. Conway. *Merchants of Doubt.* New York: Bloomsbury Press, 2010.

Further Reading

Maslin, M. *Global Warming: A Very Short Introduction.* Oxford: Oxford University Press, 2004.

Randers, J. *A Global Forecast for the Next Forty Years.* White River Junction, VT: Chelsea Green, 2012.

Weart, S. R. *The Discovery of Global Warming.* Cambridge, MA: Harvard University Press, 2003.

Index

Acid Deposition Act, 120
acid rain, 51, 114, 117–118; in China, 121; from coal-fired power plants, 119; in India, 85; legislation on, 121; natural sources of, 120
acquired immunodeficiency syndrome (AIDS), 92
activated sludge, 39
aerobic life forms, 40
African American women, 179
agriculture, 8
Air Pollution Control Act, 131
Air Quality Act, 131
Air Quality Agreement, 137
alcohol, 28
algal colonies, 51
Altman, Sidney, 59
alum, water treatment with, 18, 19
American bald eagle, 81
American Chemistry Council (ACC), 99–100
American exceptionalism, 137
American Geophysical Union to the Society for Conservation Biology, 182

ammonium nitrate, 47
amoxicillin, 92
anterior pituitary gland, 90
antibiotics, 44
anti-knocking compound, 64
aquatic animals, 47
aquatic environment, 43
aquatic life forms, 31
arctic ice, 150
Arendt, Hannah, 184
aromatic rings, 83–84
Arrhenius, Svante, 153
arsenic pollution, 12, 54
Associated Press, 113
atmospheric stability, 151
automotive technologies, 144
Axiall Corp, 70

bacteria, 11–12
Baltic Sea, 52
Beecher, Henry Ward, 129
benzene, 23, 83, *83*
biochemical oxygen demand (BOD), 32, 37, 38
biochemistry, 90–91

bioconcentration, 73, 87; analogy
for, 88; of DDT, 87, 88; predictions
on, 98
biological waste, oxidization of, 32
birds, estrogens in, 89
bisphenol A (BPA), 94, 98, 99
Blair, Eric Arthur, 140
bleach, 19
blood–brain barrier, 62
blue-green algae (cyanobacteria), 48
Borenstein, Seth, 113
bottled drinking water: cost of, 22;
price of, 24; U.S. citizens spending
on, 28; waste from, 22–23
Brattain, Walter Houser, 143
British Sanitary Act of 1845, 36
Bush, George W., 2, 76, 77, 121, 145;
environmental protection opposed
by, 135; science delegitimized by,
135

cancer: bisphenol A link to, 99; causes
of, 108; chemicals not causing, 109,
110, 110; chlorinated drinking water
link to, 106; chlorine as cause of, 21;
deaths from, 104; increasing
occurrences of, 80. See also
carcinogens
cancer rate response line, 107–108,
108
capitalism, 177
carbon calculators, 176
carbon dioxide (CO₂), 115; atmospheric
levels of, 150, 154; emissions of,
131–132, 167, 168; government
monitoring of, 154; prehistoric levels
of, 155; projected levels of, 156
carbon monoxide (CO), 65
carcinogens, 20–21, 23
Carey, J., 158
Carson, Rachel, 80, 85, 89
catalytic converters, 65, 117
cations, 61
Center for Disease Control and
Prevention (CDC), 103

cesspools, 36
Chemical Abstracts Service (CAS)
Registry, 8
chemical industry, 49
Chesapeake Bay, 52
child mortality, 24–25, 26
China, 164; acid rain in, 121; carbon
dioxide emissions of, 168; coal from,
119; greenhouse gas emissions in,
165; pollution from, 57
chloralkali plants, 70, 76
chlorinated drinking water, 106, 107
chlorine: carcinogens created by, 20–21;
harm from, 83–84; large amounts
of, 37; water treatment with, 20
chlorine monoxide, 126
chlorofluorocarbons (CFCs), 124;
hydrofluorocarbons replacement of,
126; impact of, 125; levels of, 127
cholera, 11; outbreaks of, 34–35;
prevention of, 37, 40
chronic lower respiratory diseases,
104
cinnabar, 69
Citizens Climate Lobby, 183
civilizations, 30–31, 33
Clean Air Act, 76, 99, 120, 130, 164
Clean Air Interstate Rule, 120
Clean Water Act, 13, 99, 130
Climate Action Plan, 167
climate change: affects of, 152; apathy
toward, 183; causes of, 160–161;
controversy of, 150; debate on, 163;
distanciation of, 180; fear of, 181;
from fossil fuels, 153; human cause
of, 157, 158; impacts of, 161, 173,
178, 179; inaction toward, 166;
inadequate response to, 174;
individual contributions to, 176;
local changes from, 184; mitigation
of, 161–162; science of, 151; threat
of, 175; water shortages from, 8
climate denial, 157; funding of, 162;
implicatory denial of, 181. See also
flat Earth society

CNN, 159
coal: concern for, 76; hidden costs of, 54; producers of, 119; reevaluation of, 78; use of, 71
coal-fired power plants, 50, 70; acid rain from, 119; emissions of, 3; mercury from, 71; pollution control technologies in, 77
codeine, 92
coffee, 28
Cohen, Stanley, 181
Coleridge, Samuel Taylor, 30
combined sewer systems, 42
combustion engines, 63
computers, 142
Conference of Parties-21 (COP21), 172, 173
constructed wetlands (CWs), 56
consumer products, 80
Conway, E. M., 158
Cooney, Phillip, 178
cost–benefit analysis, 69
Cronkite, Walter, 130
Crouch, A. C., 105
Cryptosporidium, 11, 20
Cummings, Alexander, 36
Cuyahoga River, 24
cyanobacteria. See blue-green algae
cyclohexane, 82–83, 82

dams, 47–48, 168–169
Dan River, 50, 54
Dark Ages, 33
data storage devices, 143
DDT (dichlorodiphenyltrichloroethane), 15, 80; bioconcentration of, 87, 88; chemical structure of, 85; in India, 85; long-term health problems from, 85; malaria fought with, 86; outlawing of, 85, 89; structure of, 84; during World War II, 84
DDT & The American Century (Kinkela), 86
dead zones, 52

death: common causes of, 103–104, 104, 107; from diarrheal diseases, 106; from voluntary action, 105
Deceit and Denial (Markowitz and Rosner), 158
Deepwater Horizon, 136
democracy, 181–182
Democrats, 131
Department of Defense (DOD), U.S., 14
Department of Energy (DOE), U.S., 14
Detroit (lake), 67
diabetes, 104
diarrheal diseases, 24–25, 106
diatomic oxygen (O_2), 31, 32, 122
dichlorofluoromethane, 123
dimethylmercury (CH_3-Hg-CH_3), 72
diseases, 24–25, 42, 103–104, 104, 106
disease theory, 34
dissolved oxygen, 31–32, 39, 41
DNA sequences, 91
Donne, John, 137
dose–response relationship, 108
Douglas, William O., 129
dredging, 47
drinking water, 11, 45; acquisition of, 7; arsenic pollution of, 12; electrolytes in, 16; in Global North, 24; Global South systems for, 23–24; municipal systems for, 22; systems of, 41–42; treatment of, 18–24; untreated types of, 21. See also bottled drinking water; chlorinated drinking water
drought, 25
Duke Power, 50, 54
DuPont, 125

Earth, 115, 153–154
E. coli, 12
ecological imagination, 173, 174, 186
ecological scientists, 174
ecosystem health, 42
Egypt, 35
Ehrlich, Paul, 130
Elizabeth I (queen), 35

Elk River, 53
Endangered Species Act, 132–133
Endangered Species Preservation Act, 130
endocrine disruptors, 43, 86–87; commonality of, 95; in consumer products, 80; dosages of, 93; intake of, 95, 97
End of Nature, The (McKibben), 149
Energy Independence and Security Act, 145
engine knocking, 63
environmental assessment (EA), 132
environmental impact statement (EIS), 132
environmentalism: in Europe, 137, 164; extremist types of, 92; genesis of, 129–130; history of, 2; legislative actions for, 134–135; Reagan opposition to, 135
environmental problems, 4, 140
Environmental Protection Agency (EPA), U.S., 12–13, 19, 77; budgeting for, 135; guidelines from, 74; major microbial treatment from, 21; pollution rating from, 102; risk assessment by, 106, 110–111; safety ratings from, 107; smog measures from, 65; standards of, 49
environmental risk assessment, 103
epidemiology, 11
Erie (lake), 18, 48, 49
Escherichia coli, 11
ethanol, 66
"ethyl," 64
Europe, 137, 145, 164; chloralkali plants in, 70; cholera in, 40; redevelopment of, 33–34; water treatment in, 19, 35–36
eutrophication, 48
ExxonMobil, 178
Exxon Valdez, 133

farm crops, 47
farming operations, 51

farm lands, 43
fast food, 28
Federal Insecticide, Fungicide, and Rodenticide Act, 132
Federal Water Pollution Control Act, 131
fertilizer, 43
Feynman, Richard P., 140
Fischetti, M., 160
fish, 44–45, 73
fishing industries, 53
flat Earth society, 157, 162
Flint, Michigan, 67–68
Flint River, 67
Food and Drug Administration (FDA), U.S., 23, 99
food industry, 96
food prices, 167
food-processing waste, 39–40
forest fires, 120
fossil fuel industry: controversy driven by, 150; greed of, 159–160; worries of, 178
fossil fuels: addiction to, 144; climate change from, 153; reduction of, 173
4-methylcyclohexanemethanol (MCHM), 53
Fox News, 159
fracking, 3, 81, 147
Francis (pope), 159
Frankland, Edward, 37
Frank R. Lautenberg Chemical Safety for the 21st Century Act, 138
Freedom Industries, 53–54
Freon-12, 123
Freons, 123, 124
freshwater, 8
freshwater–ocean interfaces, 51, 52

gasoline, 65, 78
General Motors, 123
genetically modified organisms (GMOs), 146
Germany, 14
germ theory, 11
Giardia, 11

Giddens, Anthony, 180, 182
glaciers, 74
Global North: dissolved oxygen in, 41; drinking water in, 24; good life in, 28; wastewater in, 42, 45; water prices in, 25; waterways in, 49
Global South: humanitarian efforts in, 54; municipal drinking water systems in, 23; refrigeration needs of, 127; sanitation in, 54; tobacco industry in, 66; water for, 7; water prices in, 25
global warming. *See* climate change
gold mining, 70
Gore, Al, 149, 162, 167
government, 2–3, 26–27, 100
Great Stink of 1858, 36
greed, 158, 159–160
greenhouse gas emissions, 141, 153; books on, 170; China curbing of, 165; lack of actions for, 156–157; reductions in, 161
gross domestic product (GDP), 8, 166; government budget contrasted with, 26–27; of top four economies, 27
Gulf of Mexico, 52–53

Haiyan (typhoon), 167
Hanford Nuclear Waste Site, 15
Harrington, Sir John, 35
Hazard Identification, 111
Hazardous Materials Transportation Act, 133
heart disease, *104*
Helior, S. S., 36
Henne, Albert, 123
Hentges, Steven, 100
hexane, 82, *82*
high-density polyethylene (HDPE), *96*
high-intensity forest fires, 174
high-soy diets, 95
Hockey Stick and the Climate War, The (Mann), 157

hormones: in human blood, *94*; reactions to, 91; regulation of, 90; in wastewater, 44; from wastewater plants, 97
human actions, 174
human blood, *94*
human disease transmission, 42
human endocrine system, 90
Human Genome Project, 91
human health problems, 90, 104
humanitarian efforts, 54
human life, 113
human rights, 24
hydrofluorocarbons (HFCs), 126; levels of, *127*; phasing out of, 128
hydrogen, 9
hydrogen-fuel-cell prototype engines, 144
hypothalamus, 90
hypoxia, 48

ice cores, historic record from, 155
Ideal Gas Law, 31
India, 71, 166; acid rain in, 85; carbon dioxide emissions of, 168; DDT in, 85
indicator organisms, 12
indigenous communities, 179–180
Industrial Revolution, 56–57, 74
industry, 13, 100; for chemicals, 49; decades of abuse by, 2; for fossil fuels, 150, 157–160, 178; for tobacco, 66, 104
infrared (IR) radiation, 124
Inga-Shaba Extra High Voltage DC power transmission, 55
insecticide, 85–86
insect neurons, 84
insurance, 103
Intergovernmental Panel on Climate Control (IPCC), 150, 160; conclusions of, 162; Fifth Assessment Report from, 167; impact of, 172
International Social Science Council, 175
ionic mercury (Hg^{2+}), 69

Karuk Tribe, 186
Katrina (hurricane), 178
Keeling data set, *154*, 155
Keller, George M., 140
Kerr-McGee, 136
Kettering and Midgley, 64
Kingston Fossil Plant, 50
Kinkela, David, 86
Kittman, Jamie, 64, 158
Klein, Naomi, 185
Koch, Robert, 35
Kyoto Protocol, 138–139

Charles (lake), 70
lake pollution, 48–49
lead, 59, 67, 69; in gasoline, 65, 78; in
    humans, *68*; natural presence of, 61;
    toxicity of, 62; in water pipes, 63
Lead-Based Paint Poisoning Prevention
    Act (LBPPPA), 67
Lewis, Drew, 114
Lewis dot structure, *10*
liberalism, 147
light bulbs, 145, *146*
light-emitting diodes (LEDs), 70, 145
Limbaugh, Rush, 159
limiting nutrients, 47
Liverpool, 36
Lockheed, 147
Loewig, Karl, 64
Long Island Sound, 87
Lovelock, James, 170–171
low-density polyethylene (LDPE), *96*

malaria, 86
Manhattan Project, 14
Maniates, Michael, 176
Mann, Michael E., 157, 164
Markowitz, G., 158
Marx, Karl, 177
Mauna Loa Observatory, *154*
McKibben, Bill, 149
media sources, 159
medical lore, 18
medicinal side effects, 92

menstrual periods, 95
Merchants of Doubt (Oreskes and
    Conway), 158
mercuric sulfide (HgS), 69
mercury, 59, 69, 72; in fish, 73;
    reduction of, 77; sources of, 70, 71;
    in Upper Fremont Glacier, *75*
Mercury and Air Toxics Standard
    (MATS), 77
Mersey (river), 36
metals, 61, 78
methane, *82*, 155, 168
methylmercury (CH_3-Hg^+), 69, 72
methyl tertiary butyl ether (MTBE), 66
miasma, 34
microbial contamination, 11
micro-transistor (PN junctions), 143
Midgley, Thomas, Jr., 123
Mills, C. W., 174
Mills, N. Scott, 129
Minamata Convention, 77
mineral deposits, 12
Mississippi River, 52
Molina, Mario, 125
Monsanto, 97
Montreal Protocol on Substances that
    Deplete the Ozone layer, 126, 137
Moore's law, 142
Mouras, Jean-Louis, 37
MSNBC, 159
Müller, Paul, 84
Mumford, Lewis, 140

Nairobi, Kenya, 77
Nation, The, 64
National Environmental Protection
    Act, 130, 132
National Health and Nutrition
    Examination Survey (NHANES II),
    97
National Oceanic and Atmospheric
    Administration (NOAA), U.S., 151
National Pollutant Discharge
    Elimination System (NPDES), 13,
    39, 49

National Resources Defense Council, 23, 136
national security threats, 26
natural gas, 81
Navajo Nation, 136
Neoconservatism, 137
Nimbus 7, 124
nitric acid (HNO$_3$), 118
nitrogen dioxide (NO$_2$), 65
nitrogen oxide (NO), 65, 116, 117
Nixon, Richard, 67, 159
Nobel Prize, 59, 125, 162
non–cancer-causing chemicals, 109, 110, *110*
nonnatural chemicals, 91
nonphotosynthetic organisms, 48
North American Free Trade Agreement (NAFTA), 57
North American Great Lakes, 48
North American southwest, 25
Norwegian Nobel Committee, 162
nuclear power, 146–147

Oak Ridge National Laboratory, 15
Obama, Barack, 2, 3, 138, 139, 149, 167
Occupational Safety and Health Act, 132
oceans, 150, 167
Oil Pollution Act, 133
oil spills, 130, 136
Onondaga (lake), 49
open-street drainage systems, 34
Oreskes, N., 158, 164
organic chemistry, 81, 83
organic compounds, 81–82, 83
Orwell, George, 140
overeating, 104
overpopulation, 140–141
oxygen, 9; accumulation of, 115; levels of, 116; nighttime production of, 48
ozone, 20, 21; concentration of, *126*; depletion of, 4, 169; formation of, 122; hole in, 114, 124, 164; layer types of, 123; production of, 116–117; replenishment of, 127–128

Paracelsus, 92, 102
Pasteur, Louis, 35
PCBs (polychlorinated biphenyls), 15, 94
Periodic Table of Elements, *58*, 59, 60
Perrier mineral water, 23
persistent organic pollutants (POPs), 84, 89, 97
pesticides, 81
petroleum hydrocarbons, 23
pharmaceuticals, 43, 44
phosphate, 47
phosphate pollution, 42–43
photosynthetic organisms, 115
phthalate esters, 94
phytoestrogens, 94, 95
plastics, 95, *96*
Pliny the Elder, 62
political ideology, 159
pollution: Chinese production of, 57; control technologies for, 77; decades of, 40; Environmental Protection Agency ratings of, 102; fake controls for, 136; in human blood, *94*; human health problems caused by, 90; industrial sources of, 104; nonpoint sources of, 51; with phosphate, 42–43; point sources of, 50; proteins disrupted by, 92; Russia compared to U.S. in, 15. *See also* lake pollution; river pollution
pollution-abatement equipment, 121
polycarbonate (PC), *96*
polyethylene terephthalate (PET), *96*
polypropylene (PP), *96*
polystyrene (PS), *96*
polyvinyl chloride (PVC), *96*
Population Bomb, The (Ehrlich), 130
privatization, 28
proteins, 91, 92
protest cycles, 185
public apathy, 182
public health, 3
Public Health Service, U.S., 64

radioactive wastes, 14
radioactivity, 71
rare earth metals, 145
Reagan, Ronald, 2, 135
Reed, Ron, 186
refrigeration, 127
Registration, Evaluation, Authorisation
    and Restriction of Chemicals
    (REACH), 133, 138
Reinitzer, Friedrick, 143
religious ideologies, 159
remediation: in China, 164; of lake
    pollution, 49; of wastewater, 33
renewable energy, 168, 169
Republicans, 131
Residential Lead-Based Paint Hazard
    Reduction Act, 67
Resource Conservation and Recovery
    Act, 132
risk assessment: by Environmental
    Protection Agency, 106, 110–111;
    guidelines for, 105; history of,
    102–103; limitations on, 112–113;
    for non–cancer-causing chemicals,
    109. See also environmental risk
    assessment
river pollution, 46, 50
rivers: cleanness of, 31; dams
    interference with, 47–48; industry
    near, 13; resuspension of sediments
    in, 46–47
Roman Empire, 33, 64
Rosen, C. M., 56, 57
Rosner, D., 158
rotifers, 72
Rough Ride to the Future, A (Lovelock),
    171
Rowland, Frank, 125
Russia, 15

Safe Drinking Water Act, 13, 23
Sagan, Carl, 114
Sandy (hurricane), 167
sanitation: access to, 25; development
    of, 36; fees for, 45; in Global South,

54; poor quality of, 35; progress
    in, 40
Sanskrit, 18, 19, 38
Saxe, John, 138
Schroedinger wave function equation,
    59
science: Bush delegitimization of,
    135; religious ideologies conflict
    with, 159
science communication, 182
Scientific Integrity in Policymaking,
    136
Sea Level Rise Viewer, 151
sea levels, 150
"Secret History of Lead: Special Report,
    The" (Kittman), 64
selenium, 93
selenosis, 93
septic tanks, 37
settling tanks, 37, 38
sewage: disposal of, 34, 36; domestic
    treatment of, 40; settling tanks
    treatment of, 37; treatment cost
    of, 46
sewer systems: development of, 37; of
    Liverpool, 36. See also combined
    sewer systems
shellfish, 74
Sierra Club, 135
Silent Spring (Carson), 80, 85, 89
Smith, Robert, 120
smog, 65, 116, 117
Snow, John, 11, 35
social change, 141–142, 185
social conditions, 175
social inequality, 180
social strife, 130
sociological imagination, 174, 186
sodium hypochlorite (NaClO), 19
solar energy, 168
solar system, 115
Solid Waste Disposal Act, 132
Standard Oil, 78
Steinbeck, John, 114
storage lagoons, 50

storms, 47
straight-chain compounds, 82–83
Strait of Juan de Fuca, 39
Strategic Approach to International
    Chemicals Management (SAICM),
    138
sulfur, 119
sulfur dioxide (SO$_2$), 118
sulfuric acid (H$_2$SO$_4$), 118
sulfur trioxide (SO$_3$), 118
Sun, 151
superbugs, 45
Superfund Remedial Investigations—
    Feasibility Studies, 112
Superfund sites, 14
Supreme Court, U.S., 77
Sustainable Development Goals, 28
synergistic effects, 98

tax incentives, 168
Tea Party, 137
technological advancement, 143;
    environmental challenges
    overcome by, 145; game changed
    by, 142
Teflon, 97
tetraethyl, 61
tetraethyl lead (TEL), 64, 66
This Changes Everything: Capitalism
    vs. the Climate (Naomi Klein),
    185
350.org, 183
Thwaites Glacier, 150
thyroid hormones, 93
Time, 147
tobacco industry, 66, 104
toilet, 35, 36
total ozone mapping spectrometer
    (TOMS), 124
Toxic Substances Control Act (TSCA),
    133, 138
trickling filter, 37, 38
triclosan, 45
trihalomethanes, 20
tropopause, 124

"Turning Point, The: New Hope for the
    Climate," (Gore), 167
Tyson, Neil deGrasse, 175

ultraviolet (UV), 122
ultraviolet lamps, 39
Union of Concerned Scientists, 136
United Nations (UN): International
    Panel of Climate Change established
    by, 160; successes of, 126
United Nations Environmental
    Program, 77
United Nations Millennium
    Development Goals, 26, 27
United Nations Stockholm Convention,
    85
United States (U.S.), 2; bottled drinking
    water in, 28; environmental
    performance of, 5; fracking in, 3;
    industry in, 13; pollution in, 15
University of California, 125
University of Oregon, 171, 183–184
Upper Fremont Glacier, 75
U.S. Geological Survey (USGS), 44,
    45, 73

vapor-recovery systems, 117
Victoria, B.C., 39
Vienna Convention for the Protection
    of the Ozone Layer, 126
vitamin D, 93
Vitruvius, 62
volatile hydrocarbons, 116
volcanic eruptions, 70, 74, 120
Volkswagen, 136
voluntary action, 105

wars, 167
waste stabilization ponds (WSPs), 56
wastewater: dumping of, 40; on farm
    lands, 43; in Global North, 42, 45;
    handling of, 30; in holding tanks,
    37; hormones and antibiotics in, 44;
    remediation of, 33; treatment
    facilities for, 13, 55

wastewater plants: hormones from, 97;
   privatization of, 28
water: chemical formula for, 9;
   cleanness of, 16; Lewis dot structure
   of, *10*; metals in, 61; pH level of, 118;
   properties of, 8–9; runoff of, 46;
   scarcity of, 167; shortages of, 8
waterborne diarrheal diseases,
   24–25
water pollution: definition of, 15–16;
   early days of, 9; farming operations
   causing, 51; from government,
   14–15; measurement of, 16; natural
   types of, 12; sources of, 9, 11;
   treatment of, 17–18, 144
Water Quality Act, 131

water treatment: in Europe, 19;
   inexpensive cost of, 21–22;
   weaknesses of, 23
Weber, Max, 177
Weisman, Alan, 141
Western Antarctic Ice Sheet, 150
White House Council on
   Environmental Quality, 178
Wilson, R., 105
winemaking processes, 62–63
Woodwell, G., 87
World Bank, 55
World Bank economists, 166
World War II, 14, 84, 85, 130

Yeltsin, Boris, 15